Communications
in Computer and Information Science 28

Tai-hoon Kim   Laurence T. Yang
Jong Hyuk Park   Alan Chin-Chen Chang
Thanos Vasilakos   Yan Zhang
Damien Sauveron   Xingang Wang
Young-Sik Jeong (Eds.)

Advances in Computational Science and Engineering

Second International Conference
FGCN 2008, Workshops and Symposia
Sanya, Hainan Island, China, December 13-15, 2008
Revised Selected Papers

 Springer

Volume Editors

Tai-hoon Kim
Hannam University, Daejeon, Korea
E-mail: taihoonn@empal.com

Laurence T. Yang
St. Francis Xavier University, Antigonish, Nova Scotia, Canada
E-mail: ltyang@stfx.ca

Jong Hyuk Park
Kyungnam University, Kyungnam, South Korea
E-mail: jhpark1@kyungnam.ac.kr

Alan Chin-Chen Chang
National Chung Cheng University, Chiayi County, Taiwan
E-mail: ccc@cs.ccu.edu.tw

Thanos Vasilakos
University of Western Macedonia, West Macedonia, Greece
E-mail: vasilako@ath.forthnet.gr

Yan Zhang
University of Western Sydney, Penrith South, NSW, Australia
E-mail: yan@scm.uws.edu.au

Damien Sauveron
University of Limoges, Limoges, France
E-mail: damien.sauveron@labri.fr

Xingang Wang
University of Plymouth, Plymouth, UK
E-mail: xingang.wang@plymouth.ac.uk

Young-Sik Jeong
Wonkwang University, Iksan Chonbuk, South Korea
E-mail: ysjeong@wonkwang.ac.kr

Library of Congress Control Number: Applied for

CR Subject Classification (1998): D.2, C.2, K.8, K.4, C.3

ISSN 1865-0929
ISBN 978-3-642-10237-0 Springer Berlin Heidelberg New York

springer.com

© Springer-Verlag Berlin Heidelberg 2009

Typesetting: Camera-ready by author, data conversion by Scientific Publishing Services, Chennai, India
Printed on acid-free paper   SPIN: 12792832   06/3180   5 4 3 2 1 0

Preface

As communication and networking (CN) become specialized and fragmented, it is easy to lose sight that many topics in CN have common threads and because of this, advances in one sub-discipline may transmit to another. The presentation of results between different sub-disciplines of CN encourages this interchange for the advancement of CN as a whole. Of particular interest is the hybrid approach of combining ideas from one discipline with those of another to achieve a result that is more significant than the sum of the individual parts. Through this hybrid philosophy, a new or common principle can be discovered which has the propensity to propagate throughout this multifaceted discipline.

This volume comprises the selection of extended versions of papers that were presented in their shortened form at the 2008 International Conference on Future Generation Communication and Networking (http://www.sersc.org/FGCN2008/), International Conference on Bio-Science and Bio-Technology (http://www.sersc.org/BSBT2008/), International Symposium on u- and e- Service, Science and Technology (http://www.sersc.org/UNESST2008/), International Symposium on Database Theory and Application (http://www.sersc.org/DTA2008/), International Symposium on Control and Automation (http://www.sersc.org/CA2008/), International Symposium on Signal Processing, Image Processing and Pattern Recognition (http://www.sersc.org/SIP2008/), International Symposium on Grid and Distributed Computing (ttp://www.sersc.org/GDC2008/), International Symposium on Smart Home (http://www.sersc.org/SH2008/), and 2009 Advanced Science and Technology (http://www.sersc.org/AST2009/).

We would like to acknowledge the great effort of all in the FGCN 2008, BSBT 2008, UNESST 2008, DTA 2008, CA 2008, SIP 2008, GDC 2008, SH 2008, and AST 2009 International Advisory Board and members of the International Program Committee, as well as all the organizations and individuals who supported the idea of publishing these advances in computational science and engineering, including SERSC (http://www.sersc.org/) and Springer.

We would like to give special thanks to Rosslin John Robles, Maricel O. Balitanas, Farkhod Alisherov Alisherovish, Feruza Sattarova Yusfovna. These graduate school students of Hannam University attended to the editing process of this volume with great passion.

We strongly believe in the need for continuing this undertaking in the future, in the form of a conference, journal, or book series. In this respect we welcome any feedback.

April 2009

Tai-hoon Kim
Laurence T. Yang
Jong Hyuk Park
Alan Chin-Chen Chang
Thanos Vasilakos
Yan Zhang
Damien Sauveron
Xingang Wang
Young-Sik Jeong

Organization

General Co-chairs

Luonan Chen   Osaka Sangyo University, Japan
Tughrul Arslan   Engineering and Electronics, Edinburgh University, UK

Program Co-chairs

Xiaofeng Song   Nanjing University of Aeronautics and Astronautics, China
Sang-Soo Yeo   Kyushu University, Japan

Publicity Co-chairs

Kilsoo Chun   Korea Information Security Agency, Korea
Yoojae Won   Korea Information Security Agency, Korea
Kenji Mizuguchi   National Institute of Biomedical Innovation, Japan
See-Kiong Ng   Institute for Infocomm Research, Singapore
Lusheng Wang   City University of Hong Kong
Guo-Zheng Li   Shanghai University, China
Katsuhisa Horimoto   National Institute of Advanced Industrial Science and Technology, Japan
Rodolfo Zunino   University of Genoa, Italy

International Advisory Board

Saman Kumara Halgamuge   University of Melbourne, Australia
Stephen T.C. Wong   Cornell University, USA
Suash Deb   National Institue of Science and Technology, India
Byeong-Ho KANG   University of Tasmania, Australia

Publication Chair

Yong-ik Yoon   Sookmyung Women's University, Korea

Special Session Management Co-chairs

Tai-hoon Kim   Hannam University, Korea
Wai-Chi Fang   NASA Jet Propulsion Laboratory, USA

System Management Chair

Deok-gyu Lee   ETRI, Korea

Program Committee

Ajay Kumar	The Hong Kong Polytechnic University, Hong Kong
Amar V. Singh	National Center for Computational Toxicology, USA
Andrew B.J. Teoh	Multimedia University, Malaysia
Arun Ross	West Virginia University, USA
Asai Asaithambi	University of South Dakota, USA
Carlos Juiz	Universitat de les Illes Balears, Spain
Davide Anguita	University of Genova, Italy
Dong-Yup Lee	National University of Singapore, Singaproe
Ed Harcourt	St. Lawrence University, NY, USA
Farzin Deravi	University of Kent, UK
Francisco Herrera	University of Granada, Spain
George A. Gravvanis	Democritus University of Thrace, Greece
Hale Kim	Inha University, Korea
Hamid R. Arabnia	The University of Georgia, USA
Hikmet Budak	Sabanci University, Turkey
Hujun Yin	University of Manchester, UK
Iven Mareels	University of Melbourne, Australia
Janusz Kacprzyk	Polish Academy of Sciences, Poland
Jason T. L. Wang	New Jersey Institute of Technology, USA
Javier Ortega-Garcia	Universidad Autonoma de Madrid, Spain
Jim Torresen	University of Oslo, Norway
Jin Kwak	Soonchunhyang University, Korea
Jongwook Woo	California State University, USA
Jose Alfredo Ferreira Costa	Federal University, Brazil
Jose Manuel Molina	University Carlos III of Madrid, Spain
Kevin Daimi	University of Detroit Mercy, USA
Li Xiaoli	Institute for Infocomm Research, Singapore
Liangjiang Wang	Clemson University, USA
Lusheng Wang	City University of Hong Kong, Hong Kong
Marcus Pandy	University of Melbourne, Australia
Martin Drahansky	Brno University of Technology, Czech Republic
Meena K. Sakharkar	Nanyang Technological University, Singapore
Michael E. Schuckers	St. Lawrence University, USA
Michael Wagner	University of Canberra, Australia
Pong C. Yuen	Hong Kong Baptist University, Hong Kong
QingZhong Liu	New Mexico Institute of Mining and Technology, USA
Sun-Yuan Hsieh	National Cheng Kung University, Taiwan
Syed Ali Khayam	NIIT, Pakistan
Tatsuya Akutsu	Kyoto University, Japan
Xiaohua (Tony) Hu	Drexel University, USA
Vincent Hsu	L-1 Identity Solutions, Inc., USA

Waleed Abdullah   University of Auckland, New Zealand
Witold Pedrycz   University of Alberta, Canada
Yaoqi Zhou   Indiana University of School of Informatics, USA
Yong-Nyuo Shin   KISA, Korea
Yong Shi   Kennesaw State University, USA
Yuchun Tang   Georgia State University, USA
Zhongxue Chen   Joseph Stokes Jr. Research Institute, USA
Zizhong Chen   Jacksonville State University, USA

Table of Contents

A Heuristic Approach for Designing Regional Language Based Raw–Text Extractor and Unicode Font–Mapping Tool

Debnath Bhattacharyya[1], Poulami Das[1], Debashis Ganguly[1], Kheyali Mitra[1], Swarnendu Mukherjee[1], Samir Kumar Bandyopadhyay[2], and Tai-hoon Kim[3]

[1] Computer Science and Engineering Department, Heritage Institute of Technology, Kolkata-700107, India
{debnathb,dasp88,DebashisGanguly,kheyalimitra, mukherjee.swarnendu}@gmail.com
[2] Department of Computer Science and Engineering, University of Calcutta, Kolkata-700009, India
skb1@vsnl.com
[3] Hannam University, Daejeon – 306791, Korea
taihoonn@empal.com

Abstract. Information Extraction (IE) is a type of information retrieval meant for extracting structured information. In general, the information on the web is well structured in HTML or XML format. And IE will be there to structure these documents, by using learning techniques for pattern matching in the content. A typical application of IE is to scan a set of documents written in a natural language and populate a database with the information extracted. In this paper, we have concentrated our research work to give a heuristic approach for interactive information extraction technique where the information is in Indian Regional Language. This enables any naive user to extract regional language (Indian) based document from a web document efficiently. It is just similar to a pre-programmed information extraction engine.

Keywords: Information Extraction, Indian Regional Language, Search engine, pattern matching.

1 Introduction

The internet provides the vast source of textual information at a very low cost (sometimes it is free of cost) and precisely. And that is why the World Wide Web offers a tremendously rich source of data. But it is quite unfortunate to see that it fails to satisfy a user's information needs. The only reason is that the information providers are limited in their ability to present data or information to end users. They do not have much flexibility to represent the exact data demanded by end users especially the challenge comes when extraction of information requires heterogeneous sources. Thus a new field of research related to Information Extraction (IE) incorporating a wide range of new knowledge driven applications and services have

T.-h. Kim et al. (Eds.): FGCN 2008 Workshops and Symposia, CCIS 28, pp. 1–12, 2009.

risen. An information extraction system gives opportunity to users for asking questions of documents in a database. And in reply, it answers to queries by returning possibly relevant information from the database. Any IE system is basically domain-independent as well as it has capability to build automatic subject knowledge base for its use. Not only that, it should be flexible enough to any new corpus of text and is expected to have a quick reply. Information Brokering is a process of collecting and re-distributing information. The demand for information brokering services increases as because there are very large amount of rich but unstructured data stored in the World Wide Web. That is why this field is also correlated to the IE systems. In most of the cases, it is difficult to automatically extract data from a web document. As in many cases, there is a lack of any identifiable structure of a document. However, if the structure of a document is known in advance by pattern descriptions and pattern matching, the data extraction from such documents will be easier. And using this idea, in this paper we are going to give a detailed view of how the regional language based raw texts are extracted from the web.

2 Related Works

It is highly recommended to determine whether the requirement is valid and feasible before a recommendation is reached which yields either to do nothing, improve or modify the existing system or build a new one. And that is why, before starting our actual development process we have tried to evaluate some existing methods for this particular problem. What we have found is about programming languages. As in the context of computing, it was the most burning questions how to use programming language written in Indian languages. Initially, the idea is to allow the development of computer applications in Indian languages. Nearly two decades ago, it was reported that Sanskrit is the most appropriate language for writing computer software.

Our attempt is not to write a code in regional languages, but to create a corpus for search engine. But there is no such technique or method is available. A few researches have been done on this project to study the current market trend in this field. But there are no such software tools available in the market which extract the required data (data may be in any language) from given web page accurately. Whatever software is available, can only extract the content of any HTML file but for the standard language i.e., English only. For regional language(s) of any country, the HTML to text conversion is a new approach. Not only that, even if some works related to this fields are found, they are not at all system independent.

3 Our Work

The approach which we have presented here has some goal to achieve after implementation of this technique. The main objective is to ensure the prompt, timely and efficient refinement and storage with exact font (UTF-8) of Hindi (Indian Regional Language) crude corpus as in doc file format based on some regular expressions as follows:

"<font .* face = [\'"'][a A][u U][\'"'].*>"

Not only that, it must convert the extracted raw text by reading character by character from it and writing the corresponding UTF-8 glyphs supported by the Proprietary font of that regional Language.

Now the special glyphs which are to be taken a special care here are not so easy to maintain. For that a detailed study on ISO 8859-15, (UTF8), Hindi Fonts & Proprietary Fonts [1, 2, 3] are necessary. Not only that, a detailed study of the Hindi fonts is highly required as the algorithm is based on this language [4, 5]. The approach for raw text extraction is same for any other language(s) as it follows a generic heuristic depending on regular expression searching a matched pattern of tags associated with the keyword specifying the proprietary font name as. But, here Hindi is used for our test set.

The whole technique is divided into set of modules, as discussed in details as below, which comprises the actual heuristic together upon interacting amongst each other effectively.

3.1 Main ( )

This is the main method of our algorithm. This function will take the InputFolderPath and OutputFolderPath from user and call the major module of our algorithm namely Process. It mainly represents the User-friendly interface of the algorithm.

1. Ask the absolute path for the folder containing web-pages distributed in numerous subfolders in it, representing the complete corpus website, from the user.
2. Also ask for the absolute path of the folder which will contain the extracted and font-mapped raw documents within it after successful completion of the algorithm.
3. Call PROCESS (INPUTFOLDERPATH, OUTPUTFOLDERPATH).
4. Return.

3.2 Process (Inputfolderpath, Outputfolderpath)

This method in the algorithm calls different modules like RecursiveTraverse, RawTextExtractor and FontMap.

Arguments: This function will take InputFolderPath and OutputFolderPath, as arguments and finally it will store the resultant RawTextDocuments into Output Directory.

1. It will call RecursiveTraverse module with the InputFolderPath as its argument.
2. It will store all the Web-pages, i.e., files with web extensions in list.
3. For each item in the Web-page list structure execute till Step 6.
4. Call RawTextExtractor module with the Webpage element as its argument and stores the extracted regional language raw text string.

5.   Next it will call FontMapper module with the regional language raw txt string to its argument.

6.   Now, it will store the Font Mapped Regional Language Raw Texts into Text Documents and save them into Directory structure, specified by Output-FolderPath.

3.3 Recursivetraverse (Filepath)

This function in the algorithm calls itself recursively while attempting a folder entry else call WebPageVerifier Module of the algorithm.

Arguments: This function will take the FilePath as its input and returns only Web-Pages to its invoker module.

1.   Store all the list of files and subfolder list permanently in array structure.
2.   Store the number of total elements within it.
3.   For each element of the array it will loop till Step 6.
4.   It will check whether the element is a Directory entry or not; if yes then it will call itself with the new encountered directory entry as argument.
5.   If it is not a directory entry then it will call WebPageVerifer module with the encountered file entry as the argument.
6.   If WebPageVerifier module returns true then it will return the File entry to its invoker module, i.e., Process Module.
7.   Return.

3.4 Webpageverifier (Filepath)

This function is used in the algorithm to verify whether a file entry follows specified extensions or not. So, this can be used to separate only Web-pages from all other associative site documents like image files, script files, style-sheets and many more.

Arguments: This function will take a file entry as its argument and returns True or False depending on file extension.

1.   It will calculate the position of last occurrence of '.' in the FilePath.
2.   Now from this point till end of the FilePath denotes the File Extension.
3.   It checks whether the extension extracted equals to .htm, .html or other web file extensions.
4.   If the extension is same as of possible web extensions then returns True, else returns False.

3.5 Rawtextextractor (Webpage)

This function is used to extract raw text of regional languages from the web page file specified by some regular expression and removing all the html tags.

Arguments: This function takes the Web-page file as its argument and returns the extracted raw text to its invoker module as string.

1. First fixes up a regular expression pattern which represents the general distribution of raw data in the web-page (mainly data is stored within font tags of face specifying the proprietary font name).
2. Then iteratively it searches for the regular expression pattern through the file.
3. While a match is found it retrieves the pattern and append to a string variable iteratively.
4. After this match retrieval it will find that in the string along with raw text, html tag information is also present. So, it removes all tags and separates only raw text from it.
5. It stores the Tag separated raw text into a string and returns it to its invoker module.

3.6 Fontmapper (String)

This function runs on the extracted raw text by reading character by character from it and writing the corresponding UTF-8 glyphs.

Arguments: This function will take the extracted regional language raw text as string and returns the same but after mapping it from the default proprietary character set into Unicode character set.

1. It will read character by character and executes the following operations until end of string reaches.
2. It calls iteratively the Search module with each time read decimal value of proprietary font characters as its argument.
3. It stores the decimal values of UTF-8 glyphs returned by the Search module into a string variable.
4. After completion of iteration it returns the Font-Mapped raw text string to its invoker module.

3.7 Search (Proprietaryfontvalue)

This function searches the corresponding character value to the Proprietary font glyphs in respective array structure returns the corresponding UTF-8 font value.

Arguments: This function takes the decimal value of a glyph corresponds proprietary font of regional language and returns corresponding UTF-8 font values.

1. The module runs a sequential search or linear search on the array structure representing the decimal values of the Proprietary Font glyphs of regional language.
2. While search is successful then it stores the position at where hit is found.
3. It returns the multiple decimal entries of UTF-8 font glyphs from the corresponding array structure as found at the particular position.

4 Result and Discussion

4.1 Complexity Analysis of the Stated Algorithm

The main aim of our approach is not to take into account the complexity in terms of time rather the performance of it. That means how sufficient our algorithm is to extract and stored based on the given constraints. This algorithm is tested over 97,560 files to extract the raw text in the most efficient way. The time complexity of the main modules is approximated below.

For RecursiveTraverse: This will recursively call itself until all the files are taken into account. So the time complexity will vary directly with the number of files and subfolders present in the input directory.

For RawTextExtractor: This module checks for the regular expression denoting the presence of text associated with font *AU.TTF*, i.e., written in same tag wise. Hence its complexity in terms of time as considered depends on the number of html tags present in the web Page fed as input and also while processing each tag it also consumes the time for string operations regarding regular expression analysis and hence its total time complexity comes out as to be $O(n*c)$, where 'n' corresponds to number of tags present and 'c' is constant processing time for regular expression evaluation and extraction of raw text dedicated for each tag.

For FontMapper: As this module read character by character from the extracted raw text and for each calls the Search module to find out the hexadecimal code sequence to be replaced in place as output. So the complexity of this module directly depends on the total number of character present in the extracted text and also on the number of values searched to hit the actual glyph in font-map table as in Search module. Hence, the total time complexity becomes of $O(n^2)$.

For Search: This function is dedicated to search a decimal value over a column in a two dimensional array and returns the decimal values at its corresponding column entry in linear search fashion, hence yields $O(n)$ as its time complexity where n depends on the number of glyphs present in proprietary font.

4.2 Test Results

Before analyzing the performance of the heuristics and discussing the results got from tests carried out on the input crude corpus set it is better to have some idea about what the crude corpus is meant all about.

The html files of the collected corpus, i.e., the pages from the Hindi Daily, *Amar Ujala* have been represented in Hindi text using the proprietary fonts namely *AU.TTF*, but it is expected from the end users that they may not have this proprietary font installed in their system, then in that case they cannot see the fonts and instead will see some gibberish. This erroneous display of webpage is represented in the Fig.1.

So, to overcome this deficiency and give the webpage a system independent view level flexibility and accessibility, the algorithm is carried out as refinement of crude corpus and the output of the algorithm can be displayed as the snapshot of the word document which contains UTF-8 encoded refined Hindi text as in the Fig.2.

Fig. 1. Erroneous Display of Webpage, while visited from system where the font is not installed

While processing from the Non-Unicode font glyphs to UTF-8 font glyphs, the program follows a static map table between fonts, a portion of which can be expressed as Table 1. Here, the table contains three columns, first represents the glyph from au.ttf, the second represents the hexadecimal value of the same as read in CP-1252 character encoding and the third column is the representation of the same font glyphs but with hex values of UTF-8 fonts. As, there in Unicode standard there is no glyph present for conjugate consonants, hence in the third column of the table for a single entry there can be a sequence of hex values which in combine creates the glyph same as that of in proprietary font.

मंगलवार, २२ जुलाई २००८
समाचार
चैनल्स
पंचांग

कम्युनिस्ट गणराज्य बनाने का सपना चकनाचूर
प्रचंड बोले, गठबंधन में शामिल होने से बेहतर होगा विपक्ष में बैठना
यशोदा श्रीवास्तव

काठमांडू।
राष्ट्रपति अबर उपराष्ट्रपति के चुनाव परिणाम से नेपाल को कम्युनिस्ट गणराज्य बनाने का ख्वाब देख रहे माओवादी हैरत में हैं। माओवादी प्रमुख प्रचंड ने इसे किसी दूसरे देश की साजिश करार देते हुए कहा है कि अचानक सीपीएन यूएमएल को अलग कर हमें सत्ता से दूर रखने की गहरी साजिश की गई है। प्रचंड ने यह भी कहा कि वे किसी गठबंधन की सरकार में शामिल होने से बेहतर विपक्ष में बैठना पसंद करेंगे। प्रचंड के इस बयान को नेपाली कांग्रेस के वरिष्ठ नेता अर्जुन नरसिंह केसी ने उनकी हताशा बताया है। नेपाल के प्रथम लोकतंत्र में सत्ता के दो शीर्ष पदों पर नेपाली कांग्रेस और मधेशी जनाधिकार फोरम के कब्जा जमा लेने के बाद अटकलें लगाई जा रही है कि नेपाल में गठित होने वाली संभावित सरकार भी नेपाली कांग्रेस, सीपीएन यूएमएल अबर मधेशी जनाधिकार फोरम के गठबंधन की ही होगी।
कोइराला की भूमिका को लेकर भी चर्चा
लेकिन इसके विपरीत काठमांडू के राजनीतिक गलियारों में हो रही चर्चा के मुताबिक प्रधानमंत्री कोइराला का त्याग पत्र स्वीकार करने के बाद निर्वाचित राष्ट्रपति सबसे बड़ा दल होने के नाते माओवादियों को सरकार बनाने के लिए आमंत्रित करेंगे। फिर सदन में विश्वास हासिल करने का समय देंगे। गौरतलब है कि राष्ट्रपति अबर उप राष्ट्रपति के चुनाव में माओवादियों के मतों की जो संख्या थी, वही संख्या संसद में विश्वास मत हासिल करने के दबरान भी रहेगी जो विश्वास मत के लिए अपर्याप्त है। ५९४ सदस्यों वाली संसद में बहुमत सिद्ध करने के लिए २९७ मतों की जरूरत होगी। सूत्रों के मुताबिक मतों के गणित को देखते हुए माओवादी सरकार बनाने से कन्नी काट सकते है। विश्लेषकों के मुताबिक यदि एेसा हुआ तो तीनों बड़े दलों के गठबंधन को सरकार बनाने का मबका दिया जाएगा। एेसी संस्थिति में काफी संभावना है कि सीपीएन यूएमएल नीति पर सरकार का गठन हो, जिसमें प्रधानमंत्री का पद सीपीएन यूएमएल के हिस्से में जा सकता है। नई सरकार में अंतरिम प्रधानमंत्री कोइराला की भूमिका को लेकर भी चर्चा है। खबर यह है कि कोइराला को संसद के स्पीकर पद की जिम्मेदारी सबंपी जा सकती है।
संबंधित खबरें
नेपाल में राजनीतिक समीकरण बदले
राष्ट्रपति की सुरक्षा में रहेंगे १००० जवान
राष्ट्रपति पद के लिए तीन उम्मीदवारों ने ठोकी ताल
नेपालः राष्ट्रपति चुनाव १९ को

Fig. 2. Extracted refined UTF-8 encoded Hindi text stored as Word Document

Table 1. Font Map Table for conversion from Non-Unicode proprietary font glyphs to UTF-8 font glyphs

Character	AU Hex Code	Unicode Hex Code
!	0x0021;	(int)('!');
%	0x0027;	(int)('%');
(	0x0028;	(int)('(');
)	0x0029;	(int)(')');
*	0x002A;	(int)('*');
+	0x002B;	(int)('+');
,	0x002C;	(int)(',');
–	0x002D;	(int)('-');
.	0x002E;	(int)('.');
/	0x002F;	(int)('/');
०	0x0030;	0x0966;
१	0x0031;	0x0967;
९	0x0039;	0x096F;
रु	0x004C;	0x0930; 0x0941;
रू	0x004D;	0x0930; 0x0942;
हि	0x004E;	0x0939; 0x0943;
ह्र	0x004F;	0x0939; 0x094D; 0x0930;
क्क	0x0050;	0x0915; 0x094D; 0x0915;
क्त	0x0051;	0x0915; 0x094D; 0x200D; 0x0924;
स्	0x0053;	0x0938; 0x094D; 0x200D;
ज्ञ	0x0054;	0x091E; 0x094D; 0x091C;
ख	0x00B9;	0x0916;
ख्न	0x00BA;	0x0916; 0x094D; 0x0928;
ग	0x00BB;	0x0917;
द	0x00BC;	0x0926;
ड्ट	0x00BD;	0x0921; 0x094D; 0x091F;
	0x00E4;	0x0941;
	0x00E5;	0x0942;
ा	0x00E6;	0x093E;
ि	0x00E7;	0x093F;
ी	0x00E8;	0x0940;
	0x00E9;	0x0941;
	0x00EA;	0x0942;
	0x00EB;	0x0943;

Now, if this raw text been mapped into html file then those WebPages can be viewed from anywhere independent of the system fonts installed or supported and this output can be shown as below in Fig.3.

Fig. 3. Web Page containing the refined UTF-8 encoded Hindi text

5 Conclusion

Generating a language specific search engine is a very big deal to create. Uncountable numbers of languages are there all around the world. So it is next to impossible for a person to know all of these and thus generate the data repository for them. As to prepare the data set, someone must know the flavors of respective languages. Otherwise all will be in vain. Due to this limitation, the given algorithm is not suitable for all the language but will validate all the Indian languages. Dealing with these languages is not only for search engines; it may be for generation of computing language. The most craved thing is to develop a computing language with Indian languages, so that computer programs are allowed to be written in Indian languages. Nearly two decades ago, it was reported that Sanskrit is the most appropriate language for writing computer software. So keeping pace with this valuable thoughts, there are many approaches in this field. But it is not so simple to implement. However, our attempt to prepare the refined corpus support for Information Brokering for Indian regional language based search engine is successful and it is very flexible as well as accurate to operate on.

The proposed approach is done over a static offline database of crude corpus taken down from the web site of a Hindi news paper and it comprises of more than 97 thousand files. Actually these files are nothing but daily news paper's information stored in their database since 2001. Now during the testing part of our approach with those files, we have observed that for a single file to process and refine the text within, it takes a few seconds (2 to 5 seconds based on each file contents).The extraction method what we have used here is controlling over the tree like folder structure where the crude corpus is stored. Thus it yields an automatic data extractor from any given path. During font conversion, a rigorous refinement has done over the raw data to maintain the UTF-8 encoding pattern. And this Algorithm is quite sensible for each and every letters of a sentence; especially for the combined letters. Table 1 is giving such kind of letters along with its Unicode Hexadecimal value. That is why the whole text is retrieved as well as stored properly as doc file. This makes the retrieved text more efficient for future use. These files actually makes the database for any purpose, it may be for the search engine or to generate computing languages etc. In case of generating a language specific search engine is a very big deal to create. There are uncountable numbers of languages. And hence it is next to impossible for a person to know all of these as well as generate the data repository for them. Due to this limitation, the given algorithm is not suitable for all the language but will validate all the Indian languages. However, our attempt to prepare the refined corpus support for Information Brokering for Indian regional language based search engine is successful and it is very flexible as well as accurate to operate on.

For future proposal, what we have thought is that the implemented approach can be appended to a supportive web crawler so that during the time of downloading of the corpus the whole processing can be done instantly. Using this approach, we can avoid the extra time for downloading the whole corpus and then feed it to our proposed approach. In fact, to use our approach in truly research work, the offline database support for it will not work properly. So we have to perform the task online. The idea about the usage of our approach positively is the web browsers themselves. This approach can be attached as a module to the available web browsers so that the environments where the proprietary fonts are not supported or not available, there also the content of the web browsers can be processed and converted into UTF8 fonts during the caching of the particular web page in to the client's machine. And thus the client can see regional language based contents properly. If the concept is really implemented, then our approach will get the highest honor. So being the implementer of this approach, we would be glad if any researchers who are working on the field of linguistics and like to use or implement our approach for their purpose. We hope and believe that this approach will prove its novelty only when it will be exploited properly.

References

1. http://en.wikipedia.org/wiki/ISO_8859 (visited on 12/12/2008)
2. http://aspell.net/charsets/iso8859.html (visited on 12/12/2008)
3. http://www.terena.org/activities/multiling/ml-docs/
   iso-8859.html (visited on 13/12/2008)
4. Raj, A.A., Prahallad, K.: Identification and Conversion on Font-Data in Indian Languages.
   Tech. Reports, International Institute of Information Technology,
   http://www.iiit.net/techreports/2008_1.pdf
5. Madhavi, G., Balakrishnan, M., Balakrishnan, N., Reddy, R.: Om: One tool for many (Indian) Languages. J. Zhejiang University Science 6A(11), 1348–1353 (2005),
   http://tera-3.ul.cs.cmu.edu/conference/2005/16.pdf

Aspect-Oriented Formal Specification for Real-Time Systems

Lichen Zhang

Faculty of Computer Science and Technology
Guangdong University of Technology
Guangzhou 510090, China
zhanglichen1962@163.com

Abstract. This paper presents an approach for specifying real-time systems based on aspect-oriented formal specification, which exploits the advantages and capacity of existing formal specification languages. Increasing complexity can be deal with aspect-oriented development design methods and dependability of real-time systems requires that formal development methods be applied during real-time system development cycle. The non-functional features can be separated from real-time systems based on separation of concerns, and expressed as non-functional aspects by formal methods. There is no requirement that different aspects of a system should be expressed in the same formal technique. So the different aspects can be specified by one formal specification technique or different formal specification techniques. In this paper we provide some ideas for aspect-oriented formal specification of real-time systems. Three examples illustrate the specification process of aspect-oriented formal specification for real-time systems.

Keywords: Aspect-Oriented, Real-Time Systems, Formal Specification.

1 Introduction

Aspect-oriented programming (AOP) [1] is a new software development technique, which is based on the separation of concerns. Systems could be separated into different crosscutting concerns and designed independently by using AOP techniques. Every concern is called an "aspect". Before AOP, as applications became more sophisticated, important program design decisions were difficult to capture in actual code. The implementation of the design decisions were scattered throughout, resulting in tangled code that was hard to develop and maintain. But AOP techniques can solve the problem above well, and increase comprehensibility, adaptability, and reusability of the system. AOSD model separates systems into tow parts: the core component and aspects.

When designing the system, aspects are analyzed and designed separately from the system's core functionality, such as real-time, security, error and exception handling, log, synchronization, scheduling, optimization, communication, resource sharing and distribution. After the aspects are implemented, they can be woven into the system

T.-h. Kim et al. (Eds.): FGCN 2008 Workshops and Symposia, CCIS 28, pp. 13–32, 2009.

automatically. Using AOP techniques will make the system more modular, and developers only design the aspect without considering other aspects or the core component.

Real-time feature is the most important aspect of the real-time systems, which determines the performance of the systems. So we describe the real-time feature as an independent aspect according to the AOP techniques, and design a time model to realize and manage the time aspect in order to make the real-time systems easier to design and develop and guarantee the time constraints.

The design and analysis of various types of systems, like real-time systems or communication protocols, require insight in not only the functional, but also in the real-time and performance aspects of applications involved. Research in formal methods has recognized the need for the additional support of quantitative aspects, and various initiatives have been taken to accomplish such support. In this paper, we provide some ideas for the aspect–oriented formal specification of real-time systems and three well-known case studies to validate aspect-oriented formal specification.

2 Aspect-Oriented Formal Specification

The primary goal of a software development methodology is to facilitate the creation, communication, verification and tracing of requirements, design, and implementation. To be truly effective, a modern methodology must also automatically produce implementations from designs, test cases from requirement specifications, analyses of designs and reusable component libraries. Our experience in the development of multimedia systems and research into software development methodologies have led us to conclude that existing public-domain methodologies do not permit us to achieve these goals. Most often software development methods offer excellent solutions to specific, partial aspects of system development, providing only little of help for other aspects. Classical methods for specifying and analyzing real-time systems typically focus on a limited subset of system characteristics. RTSA, for example, focuses primarily on the functionality and state-oriented behavior of real-time systems. STATEMATE provides three different graphic or diagrammatic languages for three different aspects. Module charts represent the structural aspect of the system, activity charts represent functional aspect of system, and state charts represent the behavior aspect of the system. At the other extreme, formal specification and verification methods strive for foolproof or error-free designs. They can be used to specifying and analyzing some properties such timing constraints. Thus integration of different specification methods is desired:

Integration of the methods used to specify systems requirements.

Integration of tools that support these methods.

And integration of the multiple specification fragments produced by applying these methods and tools.

In our opinion, an acceptable real-time system design methodology must synthesize the different aspects of systems, use a number of different methods and tools, consist of an orderly series of steps to assure that all aspects of the requirement and all the design aspects have been considered. The real-time system design methodology should address these problems:

Supporting specification and analysis of a system from multiple aspects to capture different perspectives on a system,

Providing methods and tools that are appropriate for different aspects for improved understandability of specifications,

Employing a formal basis to integrate multiple aspects and perform analyses with mathematical rigor,

And providing methods to handle the size and complexity required by large-scale systems.

When specifying real-time systems, decomposition and composition are the primary methods for coping with complexity. Typically, decomposition can be done in two orthogonal dimensions. First, the system is decomposed in the vertical dimension by partitioning the real-time system into loosely coupled subsystems with well-defined interfaces. Each subsystem is then decomposed further in the horizontal dimension usually in a top-down fashion. However, with this approach it is easy to introduce unnecessary complexity due to two facts. Firstly, the interfaces between subsystems are defined before capturing their collective behavior. Secondly, application-specific parts, that crosscut the two dimensions, are scattered around in different components of the subsystems. Furthermore, they are tangled with other parts. In recent years the recognition of problems related with scattering and tangling in object-oriented software systems has led to an introduction of an aspect-oriented design methodology enabling different, possibly overlapping, design concerns to be decomposed separately before composing them together. Aspect-oriented approaches use a separation of concern strategy, in which a set of simpler models, each built for a specific aspect of the system, are defined and analyzed. Each aspect model can be constructed and evolved relatively independently from other aspect models. This has three implications:

Firstly, an aspect model can focus on only one type of property, without burden of complexity from other aspects. Hence an aspect model is potentially much simpler and smaller than a traditional mixed system model. This is expected to dramatically reduce the complexity of understanding, change, and analysis.

Secondly, different levels of detail or abstraction can be used in the construction of different aspect models. This allows us to leverage of existing understanding of certain aspect of the system to reduce the complexity of modeling and analysis. For example, if the timing property of a component/subsystem is well understood, we can build an extremely simple timing model for the component.

Lastly, existing formal notations normally are suitable for describing one or a few types of system properties. By adopting the aspect concept, we can select the most suitable notation to describe a given aspect. Likewise, we can select the most suitable analysis techniques to analyze a given property.

Formal Methods refers to mathematically rigorous techniques and tools for the specification, design and verification of software and hardware systems. The phrase "mathematically rigorous" means that the specifications used in formal methods are well-formed statements in a mathematical logic and that the formal verifications are rigorous deductions in that logic (i.e. each step follows from a rule of inference and hence can be checked by a mechanical process.) The value of formal methods is that they provide a means to symbolically examine the entire state space of a digital design (whether hardware or software) and establish a correctness or safety property that is true for all possible inputs. Although the use of mathematical logic is a unifying theme across the discipline of formal methods, there is no single best "formal method". Each application domain requires different modeling methods and different proof approaches. Furthermore, even within a particular application domain, different phases of the life cycle may be best served by different tools and techniques. For example a theorem prover might be best used to analyze the correctness of a RTL [21] level description of a Fast Fourier Transform circuit, whereas algebraic derivational methods [8][22] might best be used to analyze the correctness of the design refinements into a gate-level design. Therefore there are a large number of formal methods under development throughout the world. Formal methods in software draw upon some advanced mathematics. Mathematical topics of interest include formal logic, set theory, formal languages, and automata. Formal methods support precise and rigorous specifications of those aspects of a computer system capable of being expressed in the available formal languages. Since defining what a system should do, and understanding the implications of these decisions, is the most troublesome problems in software engineering, this use of formal methods has major benefits.

Many of the existing models used to specify functional system behavior – operational techniques such as automata [23], Petri nets [12], process algebra [[26], and descriptive techniques like logics [25] – have been enriched by means to express non-functional real-time properties. Some work of is based on variants of timed automata.[23] Time restrictions are introduced by labeling transitions or states of extended finite state machines with time limits, clocks and time variables. An upper bound u and a lower bound l is assigned to each transition of the timed automaton. Once enabled, the transition may be executed not sooner than l and not later than u time units after the enabling. Similar conditions for the enabling of transitions can be formulated referring to the values of time variables. These are model-theoretic techniques in that they distinguish all execution sequences of a system into those that satisfy the timing constraints the ''good'' ones, and those that do not satisfy them the ''bad'' ones. Only those systems that can only reveal ''good'' execution sequences implement the specification. Similar timed extensions have been defined for Petri Nets, Time Petri Nets, Object Composition Petri Nets and Time Stream Petri Nets. There have also been numerous real-time extensions to process algebras, a process algebra-based QoS specification language based on the FDT LOTOS has been proposed. Temporal logics are the descriptive counterpart to specifying state transition systems by automata. Temporal logics specify qualitative temporal relationships between states. A program satisfies temporal logic specifications if all of its execution state sequences satisfy these temporal relations. As can be expected, extensions have been introduced to augment temporal logics with constructs that specify quantitative real-time relations between states. Examples are Metric Temporal Logic MTL and Quality of Service Temporal Logic QTL [11] that use real-time interval annotations to the temporal operators, and techniques introducing explicit timer variables as in Real-Time Temporal Logic RTTL or Temporal Logic of Actions TLA. LOTOS is used to specify the functional requirements and multimedia systems. At this stage, no real-time constraints or QoS guarantees are taken into account. At a second stage, we use a temporal logic based language to describe the ordering constraints over the functional behaviors and also the time-critical QoS requirements. LOTOS has been proven to be an efficient and abstract language for the specification of concurrent and non-determinism behaviors. QTL features real-time capabilities, which are required to express real-time properties.

3   Case Studies

In case studies, we illustrate aspect-oriented specification by three examples. First example is the specification of multimedia stream, Second example is the specification of the alternating bit protocol and third example is the specification of an elevator control system that illustrates the development process including static structure, dynamic behaviors and the weaving of time aspect by integration of Informal Specification and formal Specification with aspect-oriented Approach.

3.1   Aspect-Oriented Formal Specification of Multimedia Stream

In this example, the case of a multimedia stream is presented, and a structure linking a multimedia data source and a data sink. The data source and the data sink are assumed to be communicating asynchronously over an unreliable channel. We use LOTOS to specify the abstract behaviors of multimedia stream and QTL to specify real-time aspect of multimedia stream.

LOTOS is used for specifying the abstract behaviors. We believe there are a number of advantages in using process algebra such as LOTOS for the representation of abstract behaviors. Process algebraic techniques generally feature an elegant set of operators for developing concurrent systems. Thus, succinct expressions of communicating concurrent processes can be made. Similarly, the emphasis on non-determinism encourages elegant specification and abstracts away from implementation details. Furthermore, rich and tractable mathematical models of the semantics of process algebra have been developed. In LOTOS, this model is based upon concepts of equivalence through observation of the external behaviors of a specification and is derived from the seminal work of Robin Milner . Finally, due to the standardization of LOTOS and the application oriented nature of the language a large number of support tools have been developed. It should be stressed that we use standard LOTOS and do not require any semantic alterations to model time. The LOTOS specification therefore describes the possible event orders in the system, but does not relate events to real-time. Some advantages of using standard LOTOS are that changes are not required to the standard and existing toolkits can still be used.

It feels that a temporal logic offers the most natural means for expressing requirements of distributed systems. It has been demonstrated that such logics enable the abstract expression of requirements and facilitate rigorous reasoning about these requirements. More specifically, for our application domain, a real-time temporal logic is required in order to specify the real-time requirements prevalent in distributed multimedia systems. Note that the added expressiveness of real-time temporal logic over their first order equivalents (such as the real-time logic RTL) allows us to express a wider range of requirements including liveness properties. QTL is developed for the specification of distributed multimedia system requirements. QTL is based upon Koymans' metric temporal logic, MTL, which has already been demonstrated to be highly suitable for the expression of a wide range of real-time properties in the area of real-time control. QTL is designed specifically to be compatible with LOTOS and hence LOTOS events may be used as propositions in QTL. Characteristics of the logic are that it is linear time, it uses bounded operators, it employs a discrete time domain, it uses a basic set of operations (addition by constant only) and it incorporates past-tense operators as well as the more usual future-tense operators. QTL has been defined to be suitable for use in conjunction with LOTOS. However, as well as referring to LOTOS events, there is also a need to refer to data variables in QTL formulae. These data variables are necessary in order to store additional numerical information such as the number of occurrences of a particular LOTOS event or, more generally, arbitrary functions over event occurrences and their timings[11]. The functional requirements are given by LOTOS as follows[30]:

```
specification Multimedia_stream[start, play, error]: noexit
Behavior
   start;(hide source_out, sink_in, loss in
         ((Source[source_out]|||| Sink[sink_in, play,error] )
         |[source_out,sink_in]|
         Channel[source_out,sink_in,loss]))
where
   process Source[source_out]:noexit:=
                  source_out; Source[source_out]
   endproc(*Source*)
   process Sink[sink_in, play,error]:noexit:=
                  SinkBehaviour[sink_in,play] [>error;stop
         where
            process SinkBehaviour[sink_in,play]:noexit:=
                           sink_in;play;SinkBehaviour[sink_in,play]
            endproc(*SinkBehaviour*)
   endproc(*Sink*)
   process Channel[source_out,sink_in,loss]:noexit:=
                  Source_out;((sink_in;stop (*frame is transmitted successfully*)
                  []loss;stop   (*frame is lost during transmission*)
                  )
                  |||Channel[source_out,sink_in])   (*allow another frame to be sent*)
   endproc(*Channel*)
endspec(*Stream*)
```

The above functional description is now extended with real-time assumptions. These constraints are assumed to be described with QTL formulae. Each formula is assumed to describe a local property for each component of the functional specification[30].

T1: The data source arrivals are modeled with an rate of one frame every 50 ms.

$$\Box\ (source\_out \rightarrow \bigvee_{t=0}^{50} O_{=t}(\neg(source\_out) \cup_{=50-t} source\_out))$$

T2: The next constraint states that successfully transmitted frames arrive at the data sink between 80ms and 90ms.

$$\Box\ (source\_out \rightarrow \Diamond_{[80,90]} sink\_in)$$

T3: If the arrival rate the data sink is not within [15, 20]frames per second, then an error should be reported.

$$\Box(sink\_in_i \wedge (\Box_{\leq 1000}\ sink\_in_{i-21} \vee \Box_{\leq 1000} \neg sink\_in_{i-15}) \rightarrow Oerror)$$

T4: The final property states that if play is selected then it happens exactly 5ms after a frame is received.

$$\Box(sink\_in_i \wedge (\Diamond_{\leq 1000}\ sink\_in_{i-15} \wedge \Diamond_{\leq 1000} sink\_in_{i-20} \Diamond_{\leq 1000} \neg sink\_in_{i-21}) \rightarrow \Diamond_{=5} play)$$

The real-time constraints are transformed into the event scheduler as Fig.1. According to weave model, real-time aspect is weavent into the state machine which is used to express LOTOS specification. Weavent result is expressed by Timed State Machine as Fig. 2. [30]

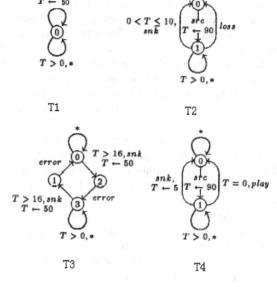

Fig. 1. Automata for Real-Time Constraints

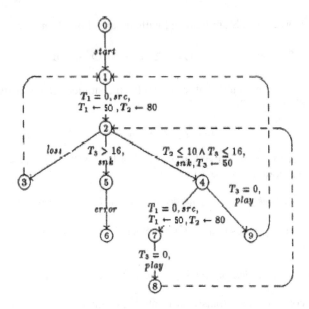

Fig. 2. Weaving Aspects into Automata

3.2 Aspect Oriented Formal Specification of ABP

The alternating bit protocol (hereafter ABP)[27][28] is a well-known protocol for communication between two parties over an unreliable loss medium (e.g. Ethernet). Briefly, each message that is sent over the medium has a single bit added, either a 1 or a 0. This bit can be used to determine if this message is a duplicate (perhaps re-sent after an acknowledgement was not received after a pre-determined amount of time). By the same token, this also allows the sender to determine if an acknowledgement just received is a duplicate (imagine the case where one message was sent, but the acknowledgement was delayed. The sender re-sends the message, and another ac-knowledgement was sent. Acknowledgement one arrives, so the next message is sent. If the delayed acknowledgement is received, the sender may otherwise assume that the message it just sent was received).

In this protocol, the loss of whole messages is possible. Corruption is possible over physical medium, but this does not form part of the protocol (the receiver can always drop a corrupted message if it is determined to be corrupted by some higher level protocol), and neither does reordering of messages. This should be reflected in the model.

For simplify the specification of ABP, the ABP protocol is divided into three aspects: the sending aspect, receiving aspect and the medium aspect, each aspect can be specified separately, finally we weave the three aspects into a single specification framework. Process algebras are a set of mathematically rigors languages with well defined operational semantics that allow for proofs of systems and propositions, either in terms of drawing an isomorphism between two systems such that they can be said to be equivalent, or in terms of deriving a system from its base terms. Process algebras CCS[[8] and PEPA[27] [29] are presented as modeling languages for concurrent systems and protocols, and are given a formal definition. PEPA is presented with a worked example of the alternating bit protocol, and it is demonstrated how equivalence relations may be drawn to prove functional and behavioral properties of this protocol, and how mathematically sound performance analysis can be applied using tools associated with this language. The methodology presented here is a generic one, and could easily be applied to other systems and may be considered a framework for modeling, proving and evaluating performance of other network protocols, as well as any concurrent and cooperating system.

One of the features of most process algebras is the facility to compose systems from specifications of individual aspects. As we will see, this is a useful feature. Within the ABP there are three distinct aspects to consider. Therefore, the modeling task should consider each of these in turn and then weave them all together to form the whole protocol.

The Sending Aspect

It is presumed that the sending aspect of the protocol will sit between some higher-level application and the physical medium. Therefore, in real life, the first thing that this part of the protocol would have to deal with is waiting for a message to be given to it for transmission. However, to simplify things, we will simulate this with a probabilistic delay. Let us assume that a model of the sender looks like[28] :

$$
\begin{aligned}
S_0 &\stackrel{def}{=} (gm, def).(send_0, def).S_1 \\
S_1 &\stackrel{def}{=} (gm, def).S_2 + (time_0, retry).S_3 + (ack_0, \top).S_5 + (ack_1, \top).S_1 \\
S_2 &\stackrel{def}{=} (time_0, retry).S_4 + (ack_0, \top).S_6 + (ack_1, \top).S_2 \\
S_3 &\stackrel{def}{=} (send_0, def).S_1 \\
S_4 &\stackrel{def}{=} (send_0, def).S_2 \\
S_5 &\stackrel{def}{=} (gm, def).(send_1, def).S_7 \\
S_6 &\stackrel{def}{=} (send_1, def).S_7 \\
S_7 &\stackrel{def}{=} (gm, def).S_8 + (time_1, retry).S_9 + (ack_0, \top).S_7 + (ack_1, \top).S_0 \\
S_8 &\stackrel{def}{=} (time_1, retry).S_9 + (ack_0, \top).S_8 + (ack_1, \top).S_3 \\
S_9 &\stackrel{def}{=} (send_1, def).S_8
\end{aligned}
$$

The Receiving Aspect

The receiving aspect is similar to the sending aspect. It is simplified in so far as there are no timeouts. Rather than accepting messages for transmission, it simply pushes the messages up-stream to the higher-level application. Let's look at the specification[28]:

$$R_0 \stackrel{def}{=} (recv_0, \mathsf{T}).(cm, def).(a_0, def).R_1$$
$$R_1 \stackrel{def}{=} (recv_0, \mathsf{T}).(a_0, def).R_1 + (recv_1, \mathsf{T}).R_2$$
$$R_2 \stackrel{def}{=} (cm, def).(a_1, def).R_3$$
$$R_3 \stackrel{def}{=} (recv_1, \mathsf{T}).(a_1, def).R_3 + (recv_0, \mathsf{T}).R_4$$
$$R_4 \stackrel{def}{=} (cm, def).(a_0, def).R_1$$

The Medium Aspect

The unreliable medium is the last aspect that requires definition. This agent will form the connection between the sending and the receiving aspects. One can infer from the definitions of these aspects that this agent must have ports labeled $\{send_0, send_1, ack_1,$ $a_0, a_1, recu_0, recu_1\}$. Also remember that this aspect has to non-deterministically loose messages. Suppose that the specification is[28] :

$$M_0 \stackrel{def}{=} (send_0, \mathsf{T}).M_1 + (send_1, \mathsf{T}).M_2 + (a_0, \mathsf{T}).M_3 + (a_1, \mathsf{T}).M_4$$
$$M_1 \stackrel{def}{=} (drop, loss).M_0 + (recv_0, def).M_0$$
$$M_2 \stackrel{def}{=} (drop, loss).M_0 + (recv_1, def).M_0$$
$$M_3 \stackrel{def}{=} (drop, loss).M_0 + (ack_0, def).M_0$$
$$M_4 \stackrel{def}{=} (drop, loss).M_0 + (ack_1, def).M_0$$

Weaving Three Aspects Together into a Single Framework

Having defined the aspects involved in the ABP, all that remains (for the modeling task) is to weave them together in order to form the larger protocol. This is done with two compositions[28] :

$$(S0 \bowtie_A M0) \bowtie_B R0$$
$$A \stackrel{def}{=} \{send_0, send_1, ack_0, ack_1\}$$
$$B \stackrel{def}{=} \{recv_0, recv_1, a_0, a_1\}$$

This defines the ABP to be a composition of S and M, interacting on channels $send_0,$ $send_1, ack_0, acka_1\}$, and a composition of M and R interacting on channels $\{recu_0,$ $recu_1, a_0, a_1\}$. Finally, ABP entire Specification is formed[28] :

$$S_0 \stackrel{def}{=} (gm, def).(send_0, def).S_1$$

$$S_1 \stackrel{def}{=} (gm, def).S_2 + (time_0, retry).S_3 + (ack_0, \mathsf{T}).S_5 + (ack_1, \mathsf{T}).S_1$$

$$S_2 \stackrel{def}{=} (time_0, retry).S_4 + (ack_0, \mathsf{T}).S_6 + (ack_1, \mathsf{T}).S_2$$

$$S_3 \stackrel{def}{=} (send_0, def).S_1$$

$$S_4 \stackrel{def}{=} (send_0, def).S_2$$

$$S_5 \stackrel{def}{=} (gm, def).(send_1, def).S_7$$

$$S_6 \stackrel{def}{=} (send_1, def).S_7$$

$$S_7 \stackrel{def}{=} (gm, def).S_8 + (time_1, retry).S_6 + (ack_0, \mathsf{T}).S_7 + (ack_1, \mathsf{T}).S_0$$

$$S_8 \stackrel{def}{=} (time_1, retry).S_9 + (ack_0, \mathsf{T}).S_8 + (ack_1, \mathsf{T}).S_3$$

$$S_9 \stackrel{def}{=} (send_1, def).S_8$$

$$R_0 \stackrel{def}{=} (recv_0, \mathsf{T}).(cm, def).(a_0, def).R_1$$

$$R_1 \stackrel{def}{=} (recv_0, \mathsf{T}).(a_0, def).R_1 + (recv_1, \mathsf{T}).R_2$$

$$R_2 \stackrel{def}{=} (cm, def).(a_1, def).R_3$$

$$R_3 \stackrel{def}{=} (recv_1, \mathsf{T}).(a_1, def).R_3 + (recv_0, \mathsf{T}).R_4$$

$$R_4 \stackrel{def}{=} (cm, def).(a_0, def).R_1$$

$$M_0 \stackrel{def}{=} (send_0, \mathsf{T}).M_1 + (send_1, \mathsf{T}).M_2 + (a_0, \mathsf{T}).M_3 + (a_1, \mathsf{T}).M_4$$

$$M_1 \stackrel{def}{=} (drop, loss).M_0 + (recv_0, def).M_0$$

$$M_2 \stackrel{def}{=} (drop, loss).M_0 + (recv_1, def).M_0$$

$$M_3 \stackrel{def}{=} (drop, loss).M_0 + (ack_0, def).M_0$$

$$M_4 \stackrel{def}{=} (drop, loss).M_0 + (ack_1, def).M_0$$

$$(S0 \bowtie_A M0) \bowtie_B R0$$

$$A \stackrel{def}{=} \{send_0, send_1, ack_0, ack_1\}$$

$$B \stackrel{def}{=} \{recv_0, recv_1, a_0, a_1\}$$

3.3 Integration of Informal Specification and Formal Specification by Aspect-Oriented Approach

UML is acquainted to be the industry-standard modeling language for the software engineering community, and it is a general purpose modeling language to be usable in a wide range of application domains. So it is very significant to research aspect-oriented real-time system modeling method based on UML[31]. However they didn't make out how to model real-time systems, and express real-time feature as an aspect. In this section, we extend the UML, and present an aspect-oriented method that model the real-time system based on UML and Real-Time Logic (RTL). Real Time Logic is a first order predicate logic invented primarily for reasoning about timing properties of real-time systems. It provides a uniform way for the specification of both relative and absolute timing of events. An elevator control system illustrates the development process including static structure, dynamic behaviors and the weaving of time aspect by integration of Informal Specification and formal Specification with aspect-oriented approach.

We consider that every floor has a pair of direction lamps indicating that the elevator is moving up or down. There is only one floor button and one direction lamp in the top floor and the bottom floor. Every floor has a sensor to monitor whether the elevator is arriving the floor.

We consider that the elevator is required to satisfy the following timing constraints[32]:

[T1] After the elevator has stopped at a particular floor, the elevator's door will open no sooner than OPEN_MIN_TIME and no later than OPEN_MAX_TIME.

[T2] After the elevator has stopped at a given floor the elevator's door will normally stay open for a STAY_OPEN_NORMAL_TIME. However, if the CloseDoorButton on board of the elevator is pressed before this timeout expires, the door will close but no sooner than STAY_OPEN_MIN_TIME.

[T3] After the door is closed, the movement of the elevator can resume, but no sooner than CLOSE_MIN_TIME, and no later than CLOSE_MAX_TIME.

Separation of Concerns from Elevator System

Several concerns can be separated from the elevator control system, such as time aspect, control aspect, and concurrency aspect. The development process of the elevator control system is shown as Fig.3. However, in this paper we only simply consider the time aspect, and will complete other aspects in our future work.

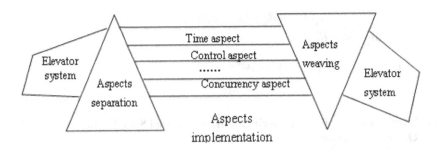

Fig. 3. The Development Process of The Elevator Control System

Structural Description Using Class Diagrams

The real-time feature of real-time systems can be modeled using UML by extending stereotypes, tagged values, and constraints before. For example, timing constraints can be added on the class to express the time feature in class diagram. But the implementation of the time feature were still scattered throughout, resulting in tangled code that was hard to develop and maintain. So we describe the real-time feature as an independent aspect according to the AOP techniques, and design a time model to realize and manage the time aspect in order to make the system easier to design and develop and guarantee the time constraints.

We separate the real-time feature as a TimeAspect, which is an instance of <<aspect>> in the elevator control system. The TimeAspect crosscuts the core functional class by stereotype <<crosscut>> in class diagram. Also timing constraints can be attached to the TimeAspect explicitly. The elevator control system class diagram is shown in Fig.4.

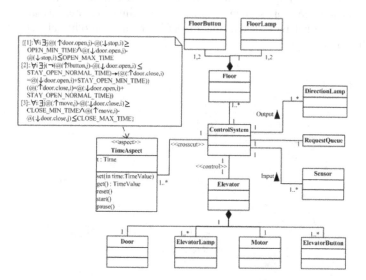

Fig. 4. The Elevator Control System Class Diagram

Behavioral Description

UML has five behavioral diagrams to describe the dynamic aspects of a system as representing its changing parts. Use case diagram organizes the behaviors of the system. Sequence diagram focuses on the time ordering of messages. Collaboration diagram emphasizes on the structural organization of objects that send and receive messages. Statechart diagram focuses on the changing state of a system driven by events. Activity diagram focuses on the flow of control from activity to activity messages. Use case diagram, collaboration diagram, and sequence diagram belong to Inter-Object behavior diagrams. While statechart diagram belong to Intra-Object behavior diagrams.

The time behavior is depicted by extending timing marks in the statechart traditionally. But we treat the time as an object (time aspect) and describe it in collaboration diagram, so as to emphasize the time behavior of the system, and refine the objects in statechart.

Collaboration Diagram

Collaboration diagram emphasizes on the structural organization of objects that send and receive messages. A collaboration diagram shows a set of objects, links among those objects, and messages sent and received by those objects. It shows classifier roles and the association roles. A classifier role is a set of features required by the collaboration. Classifier roles for core classes implement the core features required by the system. Classifier roles for aspects are services required by the core classes which are otherwise tangled with the roles of the core functional features[33]. The time aspect is time service required by the core classes in real-time systems.

The elevator control system expresses the time features as an object of TimeAspect. The behavior of the time object interacting on other objects of the system is shown in Fig. 5.

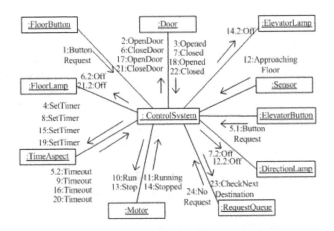

Fig. 5. Collaboration Diagram of The Elevator Control System

Statecharts

Statecharts can model the intra-object aspectual behaviors well in the object-oriented programming paradigm. However, current specification of statecharts doesn't support aspect-oriented modeling. To support aspect-orientation within the context of statecharts, we need to provide a mechanism by which the modeler can express these aspects. Statecharts modeling aspects should consider the association between aspects and transitions instead of states. Orthogonal regions, which are shown as dashed lines in statecharts, combine multiple simultaneous descriptions of the same object. The aspects can be expressed as objects, which have their own sub-states. Interactions between regions occur typically through shared variables, awareness of state changes in other regions and message passing mechanisms such as broadcasting, and propagating events [34]. The statecharts of the elevator control system is shown in Fig.6. Timing behaviors are described by the advanced features of statecharts, and the time concern is achieved implicit weaving with the core functionality of the system. Statecharts refine the model and aspects codes can be generated automatically by existing CASE tools.

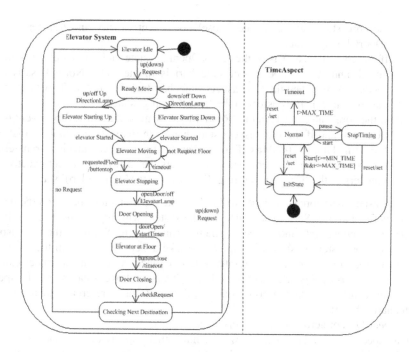

Fig. 6. Statechart of the Elevator Control System

Weaving of Time-Aspect

The time aspect can be woven into the real time system by using the UML's state-charts, as statecharts refine the model. Libraries of core and time aspect statecharts can be developed concurrently and independently, and combined only when needed for a particular application.

In this paper, we weave the time aspect with high-level declarations about how an event in the time statechart can be treated like a completely different event in the core statechart. The weaving framework permits statecharts design to be translated into skeleton code for a class. The time aspect and core statecharts objects may be joined to create orthogonal regions. In addition, the time aspect statechart can be woven by specifying which events shall be reinterpreted to have meaning in other statechart . The declarations of events reinterpretation of the elevator example can be described as follows according to the timing constraints:

[1] If the core statechart is in the 'ElevatorStopping' state and a 'openDoor' event is introduced, and if the time aspect statechart is in the state 'InitState' satisfying 't>=Min_Time&&t<=Max_Time'. Then the time aspect statechart transfers to the 'Normal' state, the core statechart treats the 'openDoor' event exactly and transfers to the 'DoorOpening' state.

[2] If the core statechart is in the 'DoorOpening' state and a 'closeDoor' event is introduced, and if the time aspect statechart is in the state 'InitState' satisfying 't>=Min_Time'. Then the time aspect statechart transfers to the 'Normal' state,

the core statechart treats the 'closeDoor' event exactly and transfers to the 'DoorClosing' state.

[3] If the core statechart is in the 'DoorOpening' state and no any event is introduced, and if the time aspect statechart is in the state 'Normal' satisfying 't>Max_Time'. Then the time aspect statechart transfers to the 'TimeOut' state, the core statechart transfers to the 'DoorClosing' state.

[4] If the core statechart is in the 'ReadyMove' state and a 'down(up)' event is introduced, and if the time aspect statechart is in the state 'InitState' satisfying 't>=Min_Time&&t<=Max_Time'. Then the time aspect statechart transfers to the 'Normal' state, the core statechart treats the 'down(up)' event exactly and transfers to the 'Elevator Starting Down(Elevator Starting Up)' state.

The time aspect and core statecharts will only make the transitions based on the declarations defined above. So it makes sure that the system will implement strictly relying on the timing constraints and guarantee the real time feature. Weaving the time aspect of the elevator system is shown in Fig. 7. We take a step forward from the work described in [35] by extending the reinterpretation function so that an aspect can be woven into other aspects or core classes. In the example above, weaving can be specified using a reference to the core 'statechart' object and a reference to the aspect 'statechart' object:

AspectID= core.crosscutBy(TimeAspect);

This specifies how an aspect is woven into other aspects or core classes. Every weaving of aspect has unique AspectID. When aspects are weaving, methods will be called to map events in the core and aspect statecharts. The declarations will hold which events need to be reinterpreted. These details will be filled in while a specific aspect is woven. The declarations above are equivalent to the expressions as follows:

[1]. reinterpretEvent(ore,"ElevatorStopping","openDoor","InitState","start/t>=OPEN_MIN_TIME&&t<=OPEN_MAX_TIME",AspectID,Statechart.PREHANDLE)

[2]. reinterpretEvent(ore,"DoorOpening","button","InitState","start/t>=STAY_OPEN_MIN_TIME",AspectID,Statechart.PREHANDLE)

[3]. reinterpretEvent(ore,"DoorOpening"," ","Normal","start/t>STAY_OPEN_NORMAL_TIME",AspectID,Statechart.PREHANDLE)

[4]. reinterpretEvent(ore,"ReadyMove","down(up)","InitState","start/t>=CLOSE_MIN_TIME&&t<= CLOSE_MAX_TIME",AspectID,Statechart.PREHANDLE).

This is done without either statechart explicitly knowing about the other. The aspect statechart can be used in any situation. Since there is no restriction on which statechart may crosscut a core statechart or it may crosscut another aspect statechart. In order to designate the real-time features explicitly, we can add the timing constraints in formal expressions like RTL to the weaving statechart as it is shown in Fig.7. The code generation will be our future work and is not discussed here.

The above three case studies have shown that aspect-oriented specification can simplify system specification. With reference to software systems concerns can be viewed as distinct system aspects or features. Concern-based decomposition has become central to software development due to the benefits that such an approach can potentially provide. A number of benefits result from maintaining a separation of

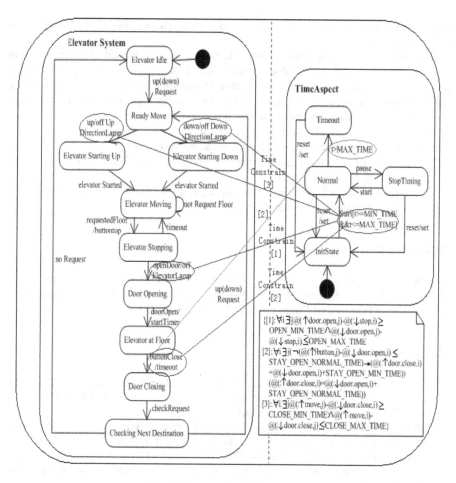

Fig. 7. Weaving the Time Aspect

concerns. Firstly, by separating out the real-time assumptions, the remaining specification of behavior is totally abstract in the sense that it does not contain any performance or implementation considerations. Secondly, the real-time assumptions are immediately identifiable and can be easily changed if necessary.

4   Conclusion and Future Research

In this paper we proposed to use aspect-oriented formal specification for real-time systems, we provided some ideas for aspect-oriented formal specification of real-time systems. Real-time systems can be divided into different aspects, each aspect can be specified independently. After each aspect has been accomplished, it can be woven back to the system. This aspect oriented formal specification method simplifies the requirement analysis process of complex multimedia systems. Three case studies were used to illustrate the specification process of aspect-oriented formal specification for real-time systems.

However aspect-oriented formal specification technique is still at initial stage at present, lots of work should be done in the future. In this paper we only provide some ideas for aspect-oriented formal specification of real-time systems. But actually many real-time systems are intricate and include a lot of other aspects that would crosscut one another. So ongoing and future research is as follows:

Integrated aspect-oriented formal specification with UML further,

Supporting multi-dimensional concerns modeling and weaving,

The semantic of aspect-oriented formal specification for real-time systems,

Automatically code generation,

Developing logic for expressing of the aspect models and the weaving of those,

Performing case studies to validate the method.

Acknowledgment

This work is supported by the National Natural Science Foundation of China under Grant No.90818008, No.60774095 and No.60474072, the Natural Science Foundation of Guangdong Province of China under Grant No.07001774 and No.04009465.

References

1. Kiczales, G., Lamping, J., Mendhekar, A., Maeda, C., Lopes, C., Loingtier, J.-M., Irving, J.: Aspect Oriented Programming. In: Aksit, M., Matsuoka, S. (eds.) ECOOP 1997. LNCS, vol. 1241, pp. 220–242. Springer, Heidelberg (1997)
2. Mohammad, R.M., Michel, R.: Separation of Concerns in the Formal Design of Real-Time Shared Data-Space Systems. In: 3rd International Conference on ACSD 2003 (2003)
3. Mousavi, Russello, M., Chaudron, G.: Using Aspect-GAMMA in the design of embedded systems. In: 7th IEEE International Conference on High-Level Design Validation and Test Workshop (2002)
4. Hoare, C.A.R.: Communicating sequential processes. Prentice-Hall, Englewood Cliffs (1985)
5. Lamport, L.: The temporal logic of actions. ACM Transactions on Programming Languages and Systems 16(3), 872–923 (1994)
6. Lynch, N.: Distributed algorithms. Morgam Kaufmann Publishers, San Francisco (1986)
7. Manna, Z., Pnueli, A.: The temporal logic of reactive and concurrent systems: specication. Springer, New York (1992)
8. Milner, R.A.: A Calculus of Communication Systems. LNCS, vol. 92. Springer, Heidelberg (1980)
9. Blair, L., Blair, G.S., Andersen, A.: Separating Functional Behaviour and Performance Constraints: Aspect-Oriented Specification, Distributed Multimedia Research Group Report MPG-98-07, Lancaster University, UK (1998)
10. Bolognesi, T., Brinksma, E.: Introduction to the ISO Specification Language LOTOS. Computer Networks and ISDN Systems 14(1), 25–59 (1987)
11. Lakas, A., Blair, G.S., Chetwynd, A.: Specification and Verification of Real-Time Properties Using LOTOS and SQTL. In: Proceedings of the 8th International Workshop on Software Specification and Design, p. 75 (1996)

12. Blair, L., Blair, G.S., Bowman, H., Chetwynd, A.G.: Formal Specification and Verification of Multimedia Systems in Open Distributed Processing. Computer Standards and Interfaces 17, 413–436 (1995)
13. Katara, M., Mikkonen, T.: Aspect-oriented specification architectures for distributed real-time systems. In: Andler, S.F., Hinchey, M.G., Offutt, J. (eds.) Proceedings of the Seventh IEEE International Conference on Engineering of Complex Computer Systems, ICECCS 2001. IEEE Computer Society Press, Los Alamitos (2001)
14. InformationTechnology E-LOTOS,
    http://lotos.site.uottawa.ca/ftp/pub/Lotos/Intro/
    FDIS15437.pdf
15. Blair, L.: Formal Specification and Verification of Distributed Multimedia Systems, PhD Thesis. Dept. of Computing, Lancaster University,Bailrigg, Lancaster, LA1 4YR, UK (1995)
16. Blair, G.S., Coulson, G., Papathomas, M., Robin, P., Stefani, J.B., Horn, F., Hazard, L.: A Programming Model and System Infrastructure for Real-Time Synchronisation in Distributed Multimedia Systems. Journal of Selected Areas in Communications (JSAC), Special Issue of Multimedia Synchronisation 14(1), 249–263 (1996)
17. Blair, G.S., Blair, L., Bowman, H., Chetwynd, A.: Formal Specification of Distributed Multimedia Systems. UCL Press (1997) (in press)
18. leduc, G.: Multimedia in the E-LOTOS process algebra. Formal methods for distributed processing: a survey of object-oriented approaches, 357–372 (2001)
19. Bowman, H., Blair, G.S., Blair, L., Chetwynd, A.G.: Time versus Abstraction in Formal Description. In: Tenney, R.L., Amer, P.D., ÜmitUyar, M. (eds.) Proceedings of the 6th International Conference on Formal Description Techniques (FORTE 1993), pp. 467–482. Elsevier Science B.V., North-Holland, IFIP (1994)
20. Bowman, H., Blair, L., Blair, G.S., Chetwynd, A.: A Formal Description Technique lSupporting Quality of Service and Media Synchronisation. In: Leopold, H., Coulson, G., Danthine, A., Hutchison, D. (eds.) COST-237 1994. LNCS, vol. 882. Springer, Heidelberg (1994)
21. Jahanian, F., Mok, A.K.: Safety Analysis of Timing Properties in Realtime Systems. IEEE Transactions on Software Engineering, 890–904 (September 1986)
22. Nicollin, X., Sifakis, J.: An Overview and Synthesis on Timed Process Algebras. In: Larsen, K.G., Skou, A. (eds.) CAV 1991. LNCS, vol. 575, pp. 526–548. Springer, Heidelberg (1992)
23. Ostroff, J.S.: Verification of Safety Critical Systems Using TTM/RTTL. In: de Bakker, J.W., Huizing, C., de Roever, W.P., Rozenberg, G. (eds.) REX 1991. LNCS, vol. 600, pp. 573–602. Springer, Heidelberg (1992)
24. Pnueli, A.: The Temporal Logic of Programs. In: Proceedings of the 18th Annual Symposium on Foundations of Computer Science, pp. 46–57 (1977)
25. Reed, G.M., Roscoe, A.W.: A Timed Model for Communicating Sequential Processes. Theoretical Computer Science 58, 249–261 (1988)
26. Regan, T.: Process Algebra for Real-Time Systems, PhD Thesis, Available from: Dept. of Computer Science, School of Cognitive and Computing Sciences,University of Sussex, Brighton, BN1 9QH (September 1991)
27. Edwards, J.: Modelling the ABP,
    http://www.cs.bris.ac.uk/~edwards/prep2001/html/node16.html
28. Bernado, M., et al.: A Stochastic Process Algebra Model for the Analysis of the Alternating Bit Protocol. In: The proceedings of the 11th International Symposium on Computer and Information Sciences (1996)

29. Clark, G., Gilmore, S., Hillston, J.: The PEPA Performance Modelling Tools. Technical Report (May 1999)
30. Lakas, A., Blair, G.S., Chetwynd, A.: A Formal Approach to the Design of QoS Parameters in Multimedia Systems,
    http://citeseer.ist.psu.edu/lakas96formal.html
31. Chavez, C., Lucena, C.J.: A Metamodel for Aspect-Oriented Modeling. In: Workshop on Aspect-Oriented Modeling with the UML. In: The First Conference on Aspect-Oriented Software Development (AOSD 2002), Netherlands (2002)
32. Dascalu, S.-M.: Combining Semi-Formal and Formal Notations in Software Specification: An Approach to Modelling Time-constrained Systems [DB/OL]. Partial fulfillment of the requirements for the degree of doctor of philosophy at Dalhousie University, Canada (September 2001)
33. Aldawud, O., Elrad, T., Bader, A.: UML Profile for Aspect-Oriented Software Development. In: Proceedings of Third International Workshop on Aspect-Oriented Modeling, Boston (2003)
34. Aldawud, O., Bader, A., Elltad, T.: Weaving with Statecharts. In: Aspect-Oriented Modeling with UML workshop at the 1st International Conference on Aspect-Oriented Software Development, Enschede (2002)
35. Mahoney, M., Bader, A., Elrad, T., Aldawud, O.: Using Aspects to Abstract and Modularize Statecharts. In: The 5th Aspect-Oriented Modeling Workshop in Conjunction with UML, Lisbon (2004)

The Advanced of Fuzzy Vault Using Password

Sumin Hong and Hwakyu Park

LIGNEX1, Yongin-City, Gyeonggi-do, Korea
fasinetcul@gmail.com
coco_kagoo@hotmail.com

Abstract. Biometric information of person is immutable and unchangeable. Thus, if it is disclosed, owner of fingerprint cannot use his fingerprint any longer. Fuzzy vault is a cryptographic framework that makes secure template storage to bind the template with a uniformly random key. In order to keep fuzzy vault secure, various schemes are studied using special data like password. K.Nandakumar proposed a scheme for hardening a fingerprint minutiae-based fuzzy vault using password. However, that scheme has vulnerabilities against several attacks. In this paper, we analyze vulnerabilities of K.Nandakumar's scheme and propose a new scheme which is secure against various attacks to fuzzy vaults.

1 Introduction

Nowadays, various internet services are provided. To use various services securely, user authentication is needed. The user authentication can be implemented in various ways, for example, using password, using smartcard, or using biometric information. In e-commerce service, for authentication, many service providers use biometric information, especially, fingerprint, because authentication using fingerprint is very simple and familiar to user. However, because biometric information is immutable and unchangeable, if biometric information is exposed, even owner cannot use his biometric information any longer. Thus, various schemes are studied to protect biometric information.

Fuzzy vault is a cryptographic framework to keep secure biometric template using encryption by a random key. Generally, fingerprint fuzzy vault consists of two factors, minutiae and chaff points. Minutiae points are unique information of fingerprint used in user authentication and chaff points are a set of points those are included in the vault to make attacker hard to extract minutiae points. However, because minutiae points of user are unique information, if attacker obtains two vaults of same user, he can extract minutiae by cross-matching. To prevent cross-matching, K.Nandakumar proposed a hardening fingerprint fuzzy vault using password. In this scheme, attacker cannot extract minutiae by cross-matching, because this scheme uses transformed minutiae points by password to make fuzzy vault. However, if an attacker obtains template before transformed and template after transformed, he can easily obtain user's password which is used in template transformation.Thus, to prevent exposure of password, in this paper, we propose a new secure fuzzy vault scheme. To solve the problem of K.Nandakumar's scheme, we use one-way hash function. Also, K.Nandakumar describes the equation

T.-h. Kim et al. (Eds.): FGCN 2008 Workshops and Symposia, CCIS 28, pp. 33–44, 2009.
© Springer-Verlag Berlin Heidelberg 2009

that transformed minutiae using plus operation between the scan image and password. We think that should just one operation. It is not only an addition but a minus, a multiplication, a division in the equation.

The remainder of this paper describes definition of fuzzy logic, fuzzy vault, fingerprint and fuzzy vault scheme for fingerprint in section 2. In section 3, we define adversarial model and analyze vulnerabilities of K.Nandakumar's scheme. Then we propose a new secure scheme. Conclusions are given in section 4.

2 Related work

2.1 Fuzzy Logic

Fuzzy is derived from fuzzy set theory to deal with reasoning that is approximate rather than precise. In binary sets with binary logic, the variables may have only 0 or 1. In fuzzy logic, the truth can range between 0 and 1 and is not constrained to the two truth values {true (1), false (0)} as in classic predicate logic[1]. Figure 1 represents logic system.

2.2 Fuzzy Vault Scheme

Fuzzy vault scheme is proposed by Ari Juels[2]. Fuzzy vault scheme is the representative method of biometric key generation and management. Figure 2 and Figure 3 represents encryption/decryption process of fuzzy vault scheme.

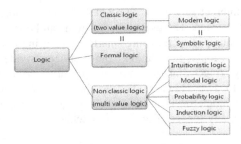

Fig. 1. Logic System

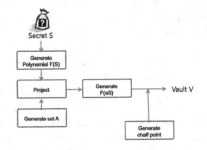

Fig. 2. Encryption of fuzzy vault scheme

To make secret S secure, secret S is encoded by the coefficients of a polynomial F, and secret S is locked using a random set of elements A. F(aS) is constructed from A and F(a) that 'a' is an element of set A. We generate randomly chaff points. Then, we generate vault V that chaff point are added to F(aS).

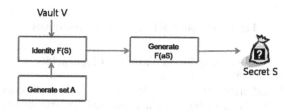

Fig. 3. Decryption of fuzzy vault scheme

We consider that want to decrypt secret S. We generate set A randomly as above. Overlapping both vault V and set A, we can obtain many points that lie on F. Using error correcting code, we can reconstruct F and obtain secret S. We describe biometric fuzzy vault next section in the concrete.

2.3 Fingerprint

Fingerprint is the process of comparing questioned and known friction skin ridge impressions(Minutiae) from fingers or palms or toes to determine if the impressions are from the same finger or palm. The flexibility of friction ridge skin means that no two finger or palm prints are ever exactly alike, even two impressions recorded immediately after each other. A known print is the intentional recording of the ridges with black printers ink rolled across a contrasting white background, typically a white card. Friction ridges can also be recorded digitally using a technique called Live-Scan. Figure 4 represents logic system.

Fig. 4. Fingerprint construction

2.4 Fuzzy Vault Scheme for Fingerprint

Umut Uludag proposed the new scheme[3]. This scheme describes authentication mechanism using biometric information which has several advantages over password-based systems. But this scheme must compute n+1 combination to make n-th degree polynomial to decrypt and so execution time of decryption process is too long.

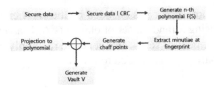

Fig. 5. Encryption of fuzzy vault for fingerprints

2.4.1 Encryption Process Using Fuzzy Vault for Fingerprints

1. Select secret value S which is 128 bits.
2. Both secure S and error correcting code (CRC, Cyclic Redundancy Check) are coherent. The CRC is parity bit checking, and used for error detection in communication channel. CRC is 16-bit code that used for error checking in decryption process. Final payload is SC = S ı C (128 + 16 = 144 bits).
3. 144 bits SC can be represented as a 9 (9=144/16) coefficients(c_8, c_7,..., c_0) of polynomial in $GF(2^{16})$. So polynomial has degree 8, $F(u) = c_8u^8 + c_7u^7 + ... + c_1u + c_0$.
4. Fingerprint template is projected to the coordinate system. Get x and y axis value of template. Extract minutiae points that have same x axis on 8th degree polynomial.
5. Generate chaff points that are not on 8th degree polynomial.
6. Minutiae points combine with chaff points to generate vault V.

2.4.2 Decryption Process Using Fuzzy Vault for Fingerprints

7. Extract minutiae points from user fingerprint at verification process. The extrac tion process is equal to the encryption process.
8. Extract the same points that have x values of the stored vault V from minutiae points of previous step.
9. For a reconstruction of n-th polynomial, make all possible combination using n+1 points.
10. Using 16 bits CRC, check error. Get secret key values S using the n-th polynomial. If an error occurs on CRC code check, we reconstruct n-th polynomial and check error using CRC again.

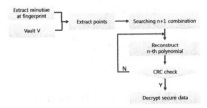

Fig. 6. Decryption of fuzzy vault for fingerprints

2.5 Hardening Fingerprint Fuzzy Vault Using Password

If biometric information is exposed on the existing fuzzy vault scheme, an immutable property of biometric information causes problem. If an attacker can compare two fuzzy vaults created using imprints of same fingerprint, he can reconstruct the polynomial using vault that has same x and y value. Therefore K.Nandakumar proposed the secure fuzzy vault scheme using minutiae points on fingerprint[4].

Figure 7 and Figure 8 show enrollment and authentication stages on the biometric authentication system of server-client structure in K.Nandakumar scheme.

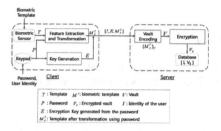

Fig. 7. Enrollment stage on fuzzy vault scheme using password

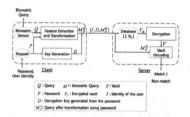

Fig. 8. Authentication stage on fuzzy vault scheme using password

Each minutiae points are represented as an element in the Galois fields $GF(2^{16})$. Let (u, v, θ) be a minutia point, where u and v indicate row and column indices in image, and θ indicate angle of the minutia point with respect to horizontal axis. The minutia attributes are uniformly and are expressed as binary strings Q_u, Q_v and Q_θ. The length of Q_u, Q_v and Q_θ are B_u, B_v and B_θ. If B_u, B_v and B_θ is 6, 5, 5, we get 16 bits number by concatenating the Q_u, Q_v and Q_θ. The polynomial P can be selected by minutiae points and these points are elements of unlocking set. At this time, K.Nandakumar uses the minutiae points transformation module using password. Assume the length of password is 64 bits(8 characters). The password is divided into 4 units of 16 bits. We classify the minutiae points into 4 classes by grouping minutiae lying in each quadrant of the image into a different class and assign one password unit to each class. We generate a permutation sequence of 4 numbers using a one way function on the password. Using this permutation sequence, we permute the 4 quadrants of the image. At this time, each quadrant is not changed. Each 16 bits password unit will be same format as a 16 bits minutia point representation. Therefore password units are divided into T_u, T_v and T_θ of length B_u, B_v and B_θ. T_u and T_v are determined by amount of translation along x axis and y axis. And T_θ. is determined by amount change in minutia point angle. New minutiae points are obtained by adding the translation value to the original values modulo appropriate range.

$$Q'_u = (Q_u + T_u) \bmod (2^{B_u})$$
$$Q'_v = (Q_v + T_v) \bmod (2^{B_v}) \text{------------------------------------} (1)$$
$$Q'_\theta = (Q_\theta + T_\theta) \bmod (2^{B_\theta})$$

3 Vulnerability Analysis of Hardening Fuzzy Vault Using Password

3.1 Adversarial Model

This section explains attacks which are considered for analysis of security for fuzzy vault using password. Figure 9 shows that attack point in fuzzy vault using password.

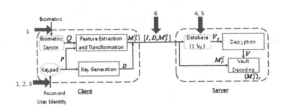

Fig. 9. Attack point in fuzzy vault using password

1. Brute Force Attack [5][6][7]
A brute force attack is that invalidates a cryptographic scheme by attempt to try on all occasions. Here the occasion means all possible keys for decrypt a message. In a general way, the theoretical possibility of a brute force attack is acknowledged.

2. Dictionary Attack[6][7][8]
A dictionary attack for searching password is that words on dictionary input one after the other. The attack is executed that not only words input to the letter but also a capital letter mixed a small letter or the word adds a number, and so on. A dictionary storing hundreds of thousands of word is treated by computer automatically.

3. Hybrid Attack[6][8]
If an attacker is fails a dictionary attack, and then he try a hybrid attack where the dictionary file is used again, but with common character substitutions or appending a character to either end of the word. By way of example, dictionary word "password" is varied as "p@ssword", "passw0rd", "pass1word", "5passw0rd". All these variations would be broken with a good cracking tool such as l0phtcrack(LC5).

4. Attacks via Record Multiplicity(ARM)[9]
We consider attacks against database in biometric security systems. In biometric security systems, database is a common target of attack. A traditional biometric system stores original template in a database, and original template used in authentication/identification. If an attacker can access to the database, all template data (T) can be compromised. In fact, an illegal access to databases is commonplace. Even if template

is encrypted, an attacker may access data in memory using a worm or virus. So we must assume that an attacker can access to the database, and retrieve any data stored in.

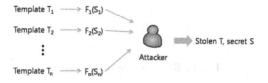

Template T_1 ⟶ $F_1(S_1)$

Template T_2 ⟶ $F_2(S_2)$

⋮

Template T_n ⟶ $F_n(S_n)$

Attacker

Stolen T, secret S

Fig. 10. Attacks via Record Multiplicity

In Figure 10, we show multiple enrollments for same biometric template T. Each enrollment has its secret S, and secret S is encoded ($F_1(S_1)$ to $F_n(S_n)$). It is transmitted and stored by different systems. An attack via record multiplicity(ARM) is that if an attacker can obtain several of these encodings, he can correlate the data contained encodings to link the databases, and may obtain some cases to directly retrieve T and $S_1 \ldots S_n$.

In fuzzy vault scheme, an encoding is called a vault. The vault is constructed by user's minutiae and chaff points. We assumed that two or more fuzzy vaults that are generated using the same fingerprint which are stored in different system. Since the minutiae points in vault generated from same fingerprint, if an attacker obtains two or more fuzzy vaults, he can obtain minutiae in vault as compare with each vault.

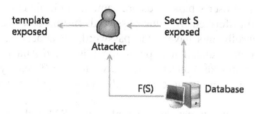

template exposed

Attacker

Secret S exposed

F(S)   Database

Fig. 11. Surreptitious Key-Inversion Attack(SKI)

5. Surreptitious Key-Inversion Attack[9]

This attack assumes that an attacker can obtain a secret of authorized user by a weak point between different modules of the system. For example, if secret S is an ID for user login or a cryptographic secret key, it may be possible to intercept secret S as submitted or used in authentication or verification system, by an insider. In case of an external attacker, traditional system is easily attacked by unprotected data transmission, unencrypted memory or has utilized Trojan horse programs to intercept data. Figure 11 shows surreptitious key-inversion attack(SKI).

In fuzzy vault scheme, secret S is added CRC(Cyclic Redundancy Check) code, then is used as coefficient of polynomial. An attacker can obtain minutiae points of authorized user by comparing with the vault and the polynomial. In this case, the vault is that minutiae point adds chaff point, and the polynomial is constructed with given coefficients.

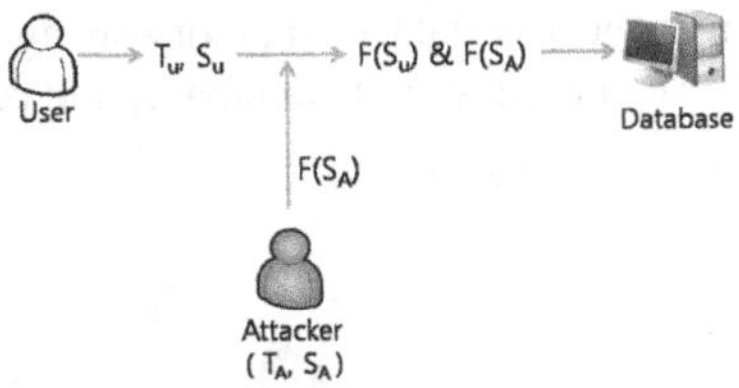

Fig. 12. Blended Substitution Attack

6. Blended Substitution Attack[9]

In a substitution attack, a user enrolls his biometric data TU and a secret SU. In user enrollment, an attacker injects another set of biometric data TA and another secret SA in the user's template. The attacker's data may be directly injected before encoding. Then database has only the attacker's data.

3.2 The Problem of K. Nandakumar's Scheme

Even if each vault is constructed with different polynomials and chaff points, if same biometric data is reused for different vaults, security of the vault decreases. If an attacker can obtain two vaults from same biometric data, he can easily identify hidden minutiae points by correlating two vaults. This vulnerability allows cross-matching of templates across different systems [9]. To solve this problem, K.Nandakumar proposed a new scheme for hardening a fingerprint minutiae-based fuzzy vault using password. However, if user's passwords are same in multiple enrollments, all fingerprints of user are transformed by same password. In this case, because fingerprints used in multiple enrollments have same password, an attacker can compute transformed template in the vaults. This type of attack is called Attack via Record Multiplicity (ARM). If an attacker collects enrolled vaults in different systems which have same implementation, then he can compute the transformed template by comparison each vault [9].

If an attacker computes transformed template in ARM and obtain biometric template of a genuine user, he can easily compute password of user.

3.3 The Attacks Scenario

1. Client enrolls transformed template by password T at different systems.
2. Each server of different systems generates vault using transformed template of client and chaff points.
3. An attacker attempts ARM to database of each sever, then can obtain many vaults V [9].
4. An attacker can collect transformed templates Q' of client by comparison each vault.
5. An attacker can obtain a fingerprint impression Q of the genuine user without any knowledge [9].
6. An attacker who has template Q and transformed template Q' of authorized client can compute password T from (1): $Q'_u = (Q_u + T_u) \bmod (2^{Bu})$, $Q'v = (Q_v + T_v) \bmod (2^{Bv})$, $Q'_\theta = (Q_\theta + T_\theta) \bmod (2^{B\theta})$. $Q' + N \cdot 2^B = Q + T$. $Q' + N \cdot 2^B - Q = T$. B is length

of Q as above. Since the Q is 16 bit, the maximum of B is 16. So the maximum of 2^{Bu} or 2^{Bv} or $2^{B\theta}$ is 2^{13}. The maximum of N is 13. This computation is not difficult to attacker.

7. Therefore an attacker's authentication can be success.

3.4 Proposed Scheme

To solve the problem of K.Nandakumar's scheme, we use one-way hash function. The original minutiae points are transformed by hashed password. In this way, an attacker cannot obtain password when user's template and transformed template are exposed. We describe 3 algorithms, template transformation, enrollment, and authentication. In template transformation process, an authorized user offers his fingerprint template to biometric sensor, and input password to keypad. Since the length of password is over 8 character, we protect password and biometric from an attack-brute force attack, dictionary attack, hybrid attack-[10]. A user's password p is transformed by hash function H. minutiae points are extracted in user's template, and then transformed by H(p). We describe the algorithm template transformation. We denote $F \leftarrow S$ to represent the secret S as a polynomial such that the secret S to be the coefficients of the polynomial. Also $TR_{H(p)}$ is a function to transform minutiae points using hash function for password.

Public parameters: A field F
Input: Parameter t, r such that $t \leq r$. A secret S F^k,
     A set $A = \{a_i\}^t_{i=1}$, where $a_i \in F$
    Output: A set T of points $\{(x_i, y_i)^t_{i=1}\}$ such that $x_i, y_i \in F$

    Algorithm. Template transformation
        $T \leftarrow \emptyset$;
        $F \leftarrow S$;
        $TR_{H(p)} \leftarrow P$;
        for i = 1 to t do
                $TR_{H(p)}(a_i, F(a_i)) \leftarrow (a_i, F(a_i))$;
                $(x_i, y_i) \leftarrow TR_{H(p)}(a_i, F(a_i))$;
                $T \leftarrow T \cup (x_i, y_i)$;
        Output T;

In enrollment process, The user sends transformed minutiae by H(p) to server. At this time, we assume that all communication channels are secure using SSL(Secure Socket Layer). So the communication between client and server is secure against blended substitution attack. Blended substitution attack is prevented by hash function. A server generates vault to add chaff points to transformed minutiae points, and then stores it. We show detail enrollment process for proposed scheme. We denote $\in U$ uniformly random selection from a set.

Public parameters: A field F
Input: Parameter t, r such that $t \leq r$. A secret S $\in F^k$,
    A set $A = \{a_i\}^t_{i=1}$, where $a_i \in F$
    Output: A set V of points $\{(x_i, y_i)^t_{i=1}\}$ such that $x_i, y_i \in F$

Algorithm. Enrollment

```
V ← ø;
for i = t+1 to r do
        x_i, y_i ∈_U F - T;
        V ← T ∪ (x_i, y_i);
Output V;
```

After enrollment, the server stores vaults in a database. At this time, since the minutiae points in vault are transformed by H(p), if an attacker has vault, he just gets transformed minutiae. Therefore, it is secure against ARM and SKI attack. In authentication process, the server decodes the vault, and then compares transmitted minutiae with stored minutiae. If both of minutiae are same, user can be authenticated. We denote (x'_i, y'_i) $(b_i, 0)$ J overlapping of J onto the x-coordinate bi. If the result of CRCcode is positive, the algorithm outputs "match". It the result of CRCcode is negative, the algorithm outputs "non-match".

Public parameters: A field F, a Cyclic Redundancy
                    Check code CRCcode.
Input: Parameter t, r such that t ≤ r. A secret S ∈ F^k,
        A set B={b_i}$^t_{i=1}$, where b_i ∈ F
Output: A secret S' ∈ F^k ∪ {'null'}

Algorithm. Authentication

```
Q ← ø;
J ← Q ∩ V
for i = 1 to t do
        (x'_i, y'_i) —(bi, 0)→ J;
        Q ← Q ∪ (x'_i, y'_i);
D ← CRCcode(Q);
if D is positive
        output "match"
else
    output "non-match"
```

To protect the user's fingerprint, we also proposed one thing. K.Nandakumar describes the equation (1) that transformed minutiae using plus operation between the scan image and password. We think that should just one operation. It is not only an addition but a minus, a multiplication, a division in the equation. If the operations apply each quadrant using arbitrary rules or criterions, the user's biometric information and password will be protected more than from attack of malicious person.First, a fingerprint of user divided into quadrant. The minutiae that extracted from the user's fingerprint transformed according to the operation by the arbitrary rules or criterions in each quadrant. Or, the minutiae could be transformed according to the operation by the arbitrary rules or criterions when fingerprint of arbitrary user be enrolled.

For example, we assume a fingerprint of user divided into quadrant. The minutiae that extracted from the user's fingerprint transformed according to the operation by the arbitrary rules or criterions in each quadrant. The scanned fingerprint of user will be extracted and correspond to the quadrant. In this time, we describe a point of reference that according to the arbitrary rules or criterions. The upper and lower parts of this point are applied difference operations and then the minutiae will be transformed.

To add to this, it applies to the top and bottom, right and left or the quadrant from a point of reference. Second, we consider the center parts of scanned fingerprint. This part is a circle that has arbitrary size in the center of scanned fingerprint. In generally, when the fingerprint is scanned, the center parts of fingerprint should be taken the high pressure. So this fingerprint will be scan correctly. A contrary concept, the edge of fingerprint will be scanned the low pressure. And then, it may be extracted less correctly than the former part. After all, the edge of fingerprint should be constructed highly the component ratio of fallible minutiae. From information security point of view, constructed minutiae from the center circle of fingerprint has more risk than the circle that constructs extractive minutiae from scanned full fingerprint when the minutiae is exposed by attacker. So, the center circle of fingerprint image will be protected user's fingerprint using the more computational complexity than among add, minus, multiplication, division.

4 Conclusion

We explained the fuzzy logic, fuzzy vault scheme, the fuzzy vault for fingerprint, fingerprint and fingerprint fuzzy vault using password for secure structure. We indicated that the scheme using password have a problem when the password is exposed and solved the problem. Also we explain the possible attack in the scheme, and analyze security of attack for proposed scheme. To solve the problem of K.Nandakumar's scheme, we use one-way hash function. Also, K.Nandakumar describes the equation that transformed minutiae using plus operation between the scan image and password. We think that should just one operation. It is not only an addition but a minus, a multiplication, a division in the equation.The identification for biometric information offers a convenience of user. But the problem still exists that biometric information has an immutable character and is not abolished. It needs to implement the transformation module that minimizes the threat of attack. Through it, the more secure scheme will be designed.

References

1. Novák, V., Perfilieva, I., Močkoř, J.: Mathematical principles of fuzzy logic. Kluwer Academic, Dordrecht (1999)
2. Juels, A., Sudan, M.: A Fuzzy Vault Scheme. In: Proceedings of IEEE International Symposium on Information Theory, Lausanne, Switzerland, p. 408 (2002)
3. Uludag, U., Pankanti, S., Jain, A.K.: Fuzzy Vault for fingerprints. In: Kanade, T., Jain, A., Ratha, N.K. (eds.) AVBPA 2005. LNCS, vol. 3546, pp. 310–319. Springer, Heidelberg (2005)
4. Nandakumar, K., Nagar, A., Jain, A.K.: Hardening Fingerprint Fuzzy Vault Using Password. In: Lee, S.-W., Li, S.Z. (eds.) ICB 2007. LNCS, vol. 4642, pp. 927–937. Springer, Heidelberg (2007)
5. Mihăilescu, P.: The fuzzy vault for fingerprints is vulnerable to brute force attack, http://arxiv.org/abs/0708.2974v1
6. Pinkas, B., Sander, T.: Securing passwords against dictionary attacks. In: Proc. 14th ACM Conf. on Computer and Communications Security (2002)

7. `http://www.lockdown.co.uk/?pg=combi&s=articles`
8. `http://blogatopic.com/2007/09/passwords-part2/`
9. Scheirer, W.J., Boult, T.E.: Cracking Fuzzy Vaults and Biometric Encryption. Univ. of Colorado at Colorado Springs, Tech. Rep. (February 2007)
10. Password Recovery Speeds,
    `http://www.lockdown.co.uk/?pg=combi&s=articles`

An Enhanced EDD QoS Scheduler for IEEE 802.11e WLAN*

Dong-Yul Lee, Sung-Ryun Kim, and Chae-Woo Lee

School of Electrical and Computer Engineering, Ajou University
San 5 Wonchon-dong Yeongtong-gu, Suwon, Korea
{dreamhunting,srkim,cwlee}@ajou.ac.kr

Abstract. The IEEE 802.11e standard was proposed to guarantee QoS in the MAC layer. It uses new HCF(Hybrid Coordination Function) protocol, which is composed of two access functions: A distributed contention-based channel access function(EDCA) providing prioritized QoS and a centralized polling-based channel access function(HCCA) providing parameterized QoS. In this paper, we propose a new scheduling algorithm to improve the performance of HCCA. The proposed scheduler guarantees QoS by changing suitably service intervals and transmission opportunities. The service schedule is based not only on the TSPEC parameters of a flow but also on the traffic queueing. Simulation results show that the proposed scheduling algorithm is superior to the reference scheduler and SETT-EDD scheduler in terms of throughput, end-to-end delay and jitter.

1 Introduction

In recent years, with increasing demands of real-time multimedia service, a lot of attention has been given to quality of service(QoS) support in wireless networks. To maintain QoS of multimedia services we need to control packet loss, delay and jitter. But it is difficult to guarantee QoS especially when the network is overloaded since traditional wireless LANs provide a best effort service only [1].

In order to guarantee the QoS for IEEE 802.11 WLAN, the IEEE 802.11 Task Group has proposed a new protocol IEEE 802.11e [2]. In IEEE 802.11e, to provide QoS a new MAC control mechanism called HCF(Hybrid Coordination Function) is defined, which replaces the role of DCF(Distributed Coordination Function) and PCF(Point Coordination Function) in 802.11. HCF is composed of EDCA(Enhanced Distributed Channel Access) which controls the channel access in contention period and HCCA(HCF Controlled Channel Access) which controls the channel access in non-contention period. HCF is backward compatible with the legacy WLAN MAC.

While PCF operates only in CP(Contention Period), HCCA of HCF operates in CFP(Contention Free Period) as well as CP. In PCF, AP(Access Point) does

---

* This work was supported by the new faculty research fund of Ajou University.

T.-h. Kim et al. (Eds.): FGCN 2008 Workshops and Symposia, CCIS 28, pp. 45–59, 2009.
© Springer-Verlag Berlin Heidelberg 2009

not restrict the amount of time which a polled node can use, which makes it difficult to guarantee QoS. Futhermore, beacon intervals can not be precisely controlled. In HCCA, AP accepts a flow only if it can guarantee QoS of both the existing and the arriving flows. It then controls the duration that a polled station can use by assigning a time limit in the poll message. The duration is called TXOP(Transmission Opportunities) [2]-[9]. In HCCA, HC(Hybrid Coordinator) located at AP allocates TXOPs through a polling mechanism. The operation of HCCA is influenced by TSPEC(Traffic Specification) of flows. Hence a scheduler is required to achieve this operation efficiently in HCCA.

In the IEEE 802.11e standard, a reference scheduler was proposed [2]. The reference scheduler is simple to implement, however, it is efficient only for the CBR(Constant Bit Rate) traffic since it assigns the same SI(Service Interval) to all stations. For VBR(Variable Bit Rate) traffic, assigning the same SI for all flows results in the bad efficiency in which the average delay, packet loss and jitter are increased drastically as the packet sizes are varied more [7]-[9].

Many schedulers have been proposed to improve the QoS performance in terms of packet loss, delay and jitter for VBR traffic [7]-[9]. For example, SETT-EDD(Scheduling based on Estimated Transmission Time-Earliest Due Date) was proposed to decrease the packet loss and the average delay by allocating TXOP and SI adaptively using a token bucket and the EDD algorithm [8]. In the scheduler, although the average delay and the packet loss are decreased compared with the reference scheduler, when bursty traffic is applied, it does not handle the traffic properly since the token bucket guarantees the average transmission only: the amount of accumulated packets in queue becomes larger as the traffic is more bursty [6]. If the scheduler takes the queue length of the flow into consideration, it may perform better. For example, ARROW(Adaptive Resource Reservation Over WLANs) allocates TXOP using the queue length information of a flow, thus it can reduce the accumulated packets in the queue [9]. However, the queue length information delivered in the packet is the queue length of the previous frame transmission Thus, it assigns TXOP which is just enough to transmit the frames remained and the frames arrived after the previous transmission: it is difficult to empty the queue. Accordingly ARROW has the same problem with SETT-EDD, too.

In this paper, we propose a new QoS scheduler for IEEE 802.11e WLAN, which can support not only CBR but also VBR with minimal queueing delay and jitter. Our scheduler allocates TXOP and SI adaptively based on TSPEC of the flow. The TXOP allocation of proposed scheduler considers not only the queue size but also the mean date rate of a flow. Then it estimates TXOP which is just large enough to empty the queue. The proposed scheduler changes SI dynamically so that it can reduce the transmission delay also: when the buffer was not empty after previous transmission, shorter SI will results in reduced delay.

This paper is organized as follows. Section 2 explains the basic operation of IEEE 802.11e and some QoS schedulers. In section 3 we present the proposed scheduler and in section 4 we show the superiority of the proposed scheduler by simulation. Finally, in section 5 we summarize the paper.

2 Related Work

2.1 IEEE 802.11e HCCA Protocol

IEEE 802.11e was proposed to guarantee QoS in Wireless LAN. It supports for QoS based on prioritized and parameterized method, and is called HCF which consists of EDCA and HCCA. The EDCA is the prioritized protocol which enhanced Distributed Coordination Function, and it provides differentiated and distributed service for QoS with access categories(AC). AC is divided into four priorities by the traffic types, and its service is independent of others and competes with others for channel occupancy.

Since HCCA controls the channel access directly, it can control the channel access more precisely than EDCA. HCCA is the parameterized protocol which enhanced Point Coordination Function, and allocates TXOP and SI to each station in accordance with its TSPEC parameters. The main TSPEC parameters are described below.

Delay Bound (D): The maximum time allowed between arrival time of MSDU at the local MAC layer and start time of MSDU at the local MAC layer.

Nominal MSDU Size (L): Nominal size of the MSDU in octets.

Minimum Physical Rate (R): Physical bit rate assumed by the scheduler for transmit time and admission control calculations, in units of bits per second.

Maximum size MSDU (M): Size of the MSDU in octets.

Mean data rate (ρ): Average bit rate for transfer of the packets, in units of bits per second.

Maximum Burst Size (MBS): Maximum size of the data burst that can be transmitted at the peak data rate, in octets.

Minimum Service Interval (mSI): Minimum time between the start of successive TXOPs allocated to the station, in units of microseconds. Given a service interval for each TSPEC (calculated as L/ρ), the mSI contained in the service schedule is equal to the smallest service interval for any TSPEC.

Maximum service interval (MSI): Maximum time allowed between the start of successive TXOPs allocated to the station, in units of microseconds.

Minimum TXOP duration (mTD): Minimum TXOP duration allocated to the station. The mTD is equal to the maximum packet transmission time for any of the stations TSPECs. The maximum packet transmission time of a TSPEC reservation i is the time required to send a packet of size M at the minimum PHY rate.

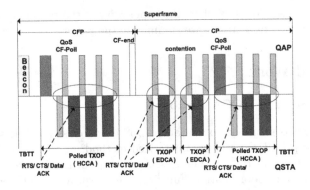

Fig. 1. The basic operation of HCCA

Maximum TXOP duration (*MTD*): Maximum TXOP duration allocated to the station. The MTD is bounded by the transmission time of the aggregate maximum burst size.

The basic operation of HCCA is as follows. First, QAP and QSTA exchanges TSPEC through traffic stream(TS) before starting the data transmission. Admission Control Unit(ACU) accepts the new flow only if can be accommodated in the available channel bandwidth without affecting QoS of the other established connections.

If a QSTA is admitted, the admitted QSTA can occupy the channel for TXOP at most. Channel assignment is repeated after SI has passed. The operation of transmission is illustrated in figure 1.

2.2   Reference Scheduler

As explained in the previous section, HCCA requires a scheduling algorithm to control the order of frame transmission, TXOPs and SI. A reference scheduler Scheduler is proposed in the IEEE 802.11e standard [2].

The basic operation of reference scheduler consists of three phases that after SI of admitted *QSTA* is calculated, TXOP is calculated from SI. Then the admission controller determines if the QSTA is admitted or not.

1st Step: SI allocation
SI is calculated as follows. SI is the greatest common divisor of the beacon interval, which is smaller than MSI of admitted QSTA. For example, let us assume that the beacon interval is 100ms and MSIs of flows(i, j, k) are, respectively, 60, 90 and 30msec and the flows are admitted in the order named as illustrated in figure 2. First, if a flow i of which MSI is 60ms is admitted, SI is chosen as 50ms because the greatest divisor for the beacon interval, which is smaller that 60ms is 50ms as illustrated in figure 2-(a). And in case of 90ms, SI is not changed because 90ms is larger than 60ms as illustrated in figure 2-(b). Finally, in case of

30ms, SI is set to 25ms since the greatest divisor for the beacon interval, which is smaller than 30ms is 25ms as illustrated in figure 2-(c).

2nd Step: TXOP allocation
HC allocates TXOP to each station in order to transmit only the average accumulated data during SI. TXOP of QSTA is calculated as follows.

$$TXOP_i = \max(\frac{N_i \times L_i}{R_i} + O, \frac{M_i}{R_i} + O) \tag{1}$$

where L_i, R_i, N_i and M_i represent, respectively, Nominal MSDU size, Physical Transmission Rate, the number of packet arrivals in SI and the maximum MSDU size of $QSTA_i$. O represents the overhead time to transmit a QoS-Poll frame, a QoS-ACK frame and IFS(Inter frame Space). N_i is calculated as follows.

$$N_i = \frac{SI \times \rho_i}{L_i} \tag{2}$$

where ρ_i denotes Mean Data Rate of $QSTA_i$.

3rd Step: Admission control algorithm
QSTA is admitted if the following inequality holds.

$$\frac{TXOP_{k+1}}{SI} + \sum_{k=1}^{k} \frac{TXOP_i}{SI} \leq \frac{T - T_{CP}}{T} \tag{3}$$

where T and T_{CP} represent the time interval for beacon and the length of contention period in the superframe, respectively.

The reference scheduler satisfies the minimum service requirements and performs well for CBR traffic. However, for VBR traffic the delay may increase and fairness can not be achieved because it uses simple round-robin polling mechanism and guarantee only the average transmission [9].

2.3   SETT-EDD Scheduler

As aforementioned, the reference scheduler guarantees the minimum performance requirement and provides satisfactory service for CBR but perform poorly for VBR traffic. SETT-EDD is a one of the representative schedulers which show improved performacnce for VBR traffic. The scheduler provides variable SI, adaptive TXOP, and earliest deadline polling mechanism [7].

The operation of SETT-EDD can be divided into two steps, i.e., TXOP allocation and SI allocation.

1st Step: TXOP allocation
SETT-EDD allocates TXOP to each station by using single token bucket. At this time, TXOP Timer replaces TXOP to give tokens to each station. Each station which received TXOP Timer reduces TXOP Timer as much as the time it used to transmit its data. If TXOP Timer remains after the transmission, it is saved

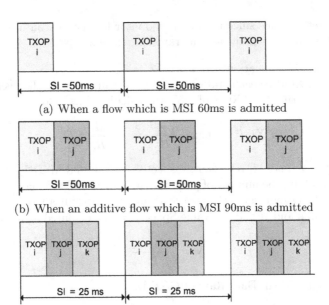

(a) When a flow which is MSI 60ms is admitted

(b) When an additive flow which is MSI 90ms is admitted

(c) When an additive flow which is MSI 30ms is admitted

Fig. 2. The Si allocation of reference scheduler

for the next transmission. The start value of TXOP for $QSTA_i$ is mTD_i which guarantees the transmission of a Maximum MSDU Size frame. The increasing rate for the token occurring in every interval is $Rate_i$, which means that TXOP Timer is increased as much as mTD in every interval to guarantee average data transmission in mSI. TXOP timer for $QSTA_i$ can increase upto MTD_i which is the time necessary to transmit a maximum burst size. $TXOP_i$ is determined between mTD_i and MTD_i of $QSTA_i$ which are calculated as follows.

$$mTD_i = \max(\frac{M_i}{R_i})$$

$$MTD_i = \frac{\Sigma MBS_i}{R_i} \qquad (4)$$

$$Rate_i = \frac{mTD_i}{mSI_i}$$

2nd Step: SI allocation
Reference scheduler is inefficient because all the station have the same SI. SI of SETT-EDD is variable because it uses work-conserving method as follows. In SETT-EDD, if the channel is idle for PIFS duration, the scheduler starts to poll the station which satisfies the following equation.

$$t_{pre}^i + mSI_i \leq t' \leq t_{pre}^i + MSI_i \qquad (5)$$

where t^i_{pre} and t' denote, respectively, the previous polled time for $QSTA_i$ and the present time. SETT-EDD chooses the QSTA which has the highest deadline calculated using the following equation.

$$Deadline_i = t^i_{pre} + MSI_i \qquad (6)$$

2.4 ARROW Scheduler

SETT-EDD can reduce packet loss and jitter in VBR because of adaptive SI and TXOP. However, as the traffic becomes more bursty, the number of packets waiting in the queue increases. This is because SETT-EDD assigns based on the average rate of QSTA. ARROW allocates TXOP considering the exact queue length of QSTA [9]. ARROW assigns TXOP based on the average rate of QSTA, but it adjust TXOP using the queue length information carried in the packet transmitted in the previous transmission. And the basic operation of SI is based on EDD algorithms like SETT-EDD. The operation of ARROW scheduler can also be divided into two steps as follows.

1st Step: TXOP allocation
TXOP assignment of ARROW is illustrated in figure 3. For simplicity only one QSTA is assumed. For x^{th} transmission of $QSTA_i$, $TXOP_i(x)$ of $QSTA_i$ is allocated at time $t_i(x)$ according to TXOP duation requested field(TDr) of the QoS control field. Here, (TDr) means the required time to transmit the packet remaining in queue of QSTA if the 4th bit of the QoS Control field is 0 when QSTA transmits the QoS data frame. Hence, if TXOP duration requested field is used when transmitting final QoS data frame in the present SI, the additional TXOP can be allocated to the next SI. Hence, QSTA can transmit the packet which is not transmitted during the current SI during the next SI.

Hence, in figure 3, ARROW first allocates $TXOP_i(x)$ at time $t_i(x)$ and the queue length information($QS_i(x)$), Queue Size of $QSTA_i$, is provided to HC with it own packet at the end of the data transmission. And next $TXOP_i(x+1)$ is assigned to $QSTA_i$ as much as $QS_i(x)$ and $QS_i(x)$ is transmitted at the end of data transmission. In the same manner, the operation is repeated. As illustrated,

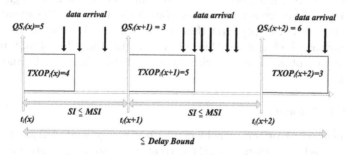

Fig. 3. The TXOP allocation method of ARROW algorithm

since using QoS Control field provides exact queue length information, the delay accumulated by bursty traffic can be decreased.

2nd Step: SI allocation

The SI allocation is as follows. As observed in figure 3, for $QSTA_i$, data arriving within the interval $[t_i(x), t_i(x+1)]$ can be sent no earlier than $TXOP_i(x+2)$. Hence, in order to keep up with delay bound, The worst case is assumed that the SI and MSI is the same and $TXOP_i(x+2) = MTD_i$, Hence the MSI of ARROW is as follows.

$$D_i \geq 2MSI_i + MTD_i \iff MSI_i \geq \frac{D_i - MTD_i}{2} \qquad (7)$$

3 Proposed Scheduling Algorithm

In this section, we present a new algorithm that makes delay, jitter and packet loss probability decrease than the existing algorithms using the characteristics of traffic.

As aforementioned,the reference scheduler presented in IEEE 802.11e is easy to implement and simple. It has low efficiency like considerable packet loss probability in VBR since SI and TXOP have fixed values. SETT-EDD, however, has a performance superior to reference scheduler in VBR by using variable TXOP and SI. But there exists accumulated packet by bursty traffic since TXOP allocation of SETT-EDD considers only average TXOP transmission and do not consider exactly bursty traffic that the characteristics cause delay increase. Also SETT-EDD has a problem which remaining packets in current SI waits unnecessarily because of fixed mSI and MSI. In order to solve the problem caused by bursty traffic, ARROW was proposed to consider queue size because the packets accumulated by bursty traffic wait in queue of QSTA. However, it is difficult for ARROW to empty queue because the queue size information is about previous transmission, and it does not overcome unnecessary wait in SETT-EDD.

Basically, the proposed algorithm is based on EDD. The proposed algorithm also uses the queue length information like ARROW. In addition the algorithm estimates the number of packets arrived after the end of the previous transmission. Then the algorithm calculates TXOP which is just large enough to clear the queue when the current transmission completes. To reduce the average delay, when the buffer is not empty after the current transmission completes, the next SI begins earlier, which can be achieved by changing the value of mSI and MSI.

The operation of proposed algorithm can be divided into TXOP and SI allocation, too.

3.1 TXOP Allocation

TXOP allocation algorithm considers the packets generated from t_{pre} when QSTA was polled at previous SI to the current time and the packets remaining in the queue at previous SI. To transmit the packets generated during SI, it needs $TXOP_{avg}^i$ which is calculated as follows.

$$TXOP^i_{avg} = \max \left[\frac{[t^i_{pre} - t'] \cdot \rho_i}{R_i} + O, \frac{M_i}{R_i} + O \right] \tag{8}$$

where t^i_{pre} and t' denote, respectively, the previous polled time for $QSTA_i$ and the present time. To transmit the packets remaining in the queue at previous SI, it needs TXOP(TDr_i) is calculated exactly the same as ARROW. Hence, the total TXOP is given as follows.

$$TXOP_i = TXOP^i_{avg} + TDr_i \tag{9}$$

Figure 4 shows the method of TXOP allocation in the proposed algorithm.

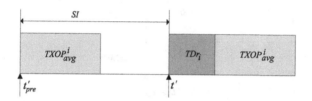

Fig. 4. TXOP assignment with proposed algorithm

3.2 SI Allocation

When the algorithm determines SI, it considers two aspects. First one is that the data generated in previous SI is delayed if the packets remain in queue as illustrated in figure 5-(a). Another is that when QSTA does not consume given TXOP, then it has longer idle time after data transmission as illustrated in figure 5-(b). In this case, the delay may become longer for subsequently arrived packets. If the next service starts earlier, the delay can be reduced.

In the first case that there are the packets waiting in queue, it causes the delay of the data arrived in previous SI as seen figure 5-(a). It is because the mSI which the QSTAs can has a authority to transmit by mSI is fixed. However, as seen in figure 5-(a), if mSI is decreased as much as TDr which is the time necessary to transmit previously queued packets, we can avoid the delay by changing the start of mSI. New mSI can be calculated as follows.

$$mSI^i_{new} = mSI_i - TDr^i_{cur} \tag{10}$$

where TDr^i_{cur} denotes the value of current TXOP Duration request.

In the second case that QSTA does not consume given TXOP as illustrated in figure 5-(b), the idle time after data transmission increases, which would lead to increased delay for the packets arrived later. If next SI starts earlier so that the idle time after data transmission remains the same, we could reduce the delay. Hence our scheduler changes mSI as follows.

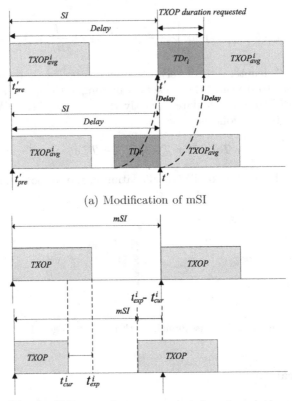

(a) Modification of mSI

(b) The difference between expected and real time when channel is idle

Fig. 5. SI allocation of the proposed scheduler

$$
\begin{aligned}
&\text{if}(t_{exp}^i - t_{cur}^i) > 0, \\
&\qquad mSI_{new}^i = mSI^i - (t_{exp}^i - t_{cur}^i)
\end{aligned}
\tag{11}
$$

$$
\begin{aligned}
&\text{otherwise,} \\
&\qquad mSI_{new}^i = mSI_i
\end{aligned}
$$

where t_{cur}^i and t_{exp}^i respectively denote the time when the data transmission completed and the expected time when $TXOP_i$ finished in previous SI. mSI_{new}^i is newly assigned mSI of $QSTA_i$.

4   Performance Evaluation of the Proposed Algorithm

In this section, in order to show the performance of our scheduler, we ran the simulation using NS-2 [12].

4.1   Simulation Scenarios

IEEE 802.11e scenario was constructed using a star type topology to analyze the performance of the proposed algorithm. QAP and QSTA were used as the components of the network and each QSTA have an up-link traffic stream to transmit data to QAP. We compared performance during 1000s by increasing the number of QSTAs from 1 to 12. Physical layer used in this simulation is IEEE 802.11b. For MAC level, IEEE 802.11e HCCA module which was developed by Pisa University in Italy [13] was used. Parameters used in the simulation are summarized in Table 1.

Table 1. Simulation Parameters

Parameters	Value
Slot Time	20s
SIFS	10s
PIFS	30s
Preamble Length	144 bits
PLCP Header Length	48 bits
PLCP TX Rate	1 Mbps
MAC Header Length	60 Bytes
Basic Tx Rate	1 Mbps
Data Rate	11 Mbps

The two simulation scenario were used. Scenario 1 used VoIP for CBR traffic and measured throughput and mean delay. Scenario 2 used MPEG4 video(Jurassic Park) for VBR traffic [14] and measured the burst traffic characteristics in VBR such as throughput rate, mean delay, jitter and packet loss ratio. Characteristic of the traffic used in the simulation are shown in Table 2.

Table 2. Parameters of TSPEC

TSPEC	CBR traffic	VBR traffic
Mean Data Rate	83Kbps	128Kbps
Peak Data Rate	83Kbps	1.7Mbps
Nominal MSDU SIze	208Bytes	1,300Bytes
Maximum MSDU Size	208Bytes	5,211Bytes
Maximum Burst Size	576Bytes	5,211Bytes
Minimum PHY Rate	11Mbps	11Mbps
Maximum Service Interval	30ms	40ms
Delay Bound	60ms	80ms

4.2   Analysis of the Simulation Results

Figure 6 shows throughput in scenario 1. For CBR traffic, the reference scheduler, SETT-EDD and our scheduler peform very well. There is no dropped packets for all three schedulers.

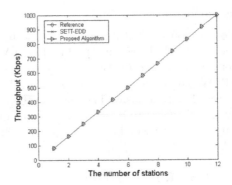

Fig. 6. Throughput in scenario 1 (CBR traffic)

Figure 7 shows the mean delay time in scenario 1. In the figure, we can see that the average delay for the reference scheduler is the highest and the delays for SETT-EDD and the proposed algorithm are the same. In the simulation, the beacon interval and the packet generation interval are set to 100ms and 30ms, respectively. Since 20ms is the greatest common divisor, the reference scheduler, allocated SI as 20ms. Packets arrived in the previous SI need to wait maximum 20ms even if there is only one STA. This is because the reference scheduler is not work-conserving. However, the other two schedulers are work conserving. Thus, if there are packets to transmit, they try to transmit the packet instead of idling the transmitter. This explains why the average delays of our scheduler and SETT-EDD are smaller than that of the reference scheduler.

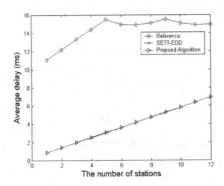

Fig. 7. Average delay in scenario 1 (CBR traffic)

Figure 8 and 9 shows the throughput and the the packet loss ratio in scenario 2 in which MPEG-4 video traffic was used. In figure 9, we can see that packet drop occurs in the reference scheduler even when the number of STAs is two. This is because the reference scheduler could not handle large burst. SETT-EDD

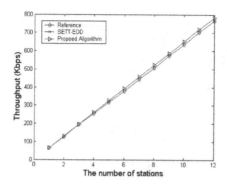

Fig. 8. Throughput in scenario 2 (VBR traffic)

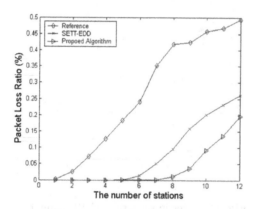

Fig. 9. Packet loss ratio in scenario 2 (VBR traffic)

began to drop packets when the number of STAs was 6, while our scheduler dropped packet when the number of STAs was 8. The proposed scheduler has better performance than SETT-EDD, because SETT-EDD uses fixed mSI and MSI and does not transmit all the packet when burst period is long. From the simulation, we can see that our algorithm can support more VBR video traffic than the reference scheduler or SETT-EDD, because even small percentage of packet drop in video traffic results in sever degradation in video quality.

Figure 10 shows the average delay in scenario 2. The delay of the proposed algorithm is the smallest among the three. This is because our scheduler aims to empty the queue after the current transmission finishes.

Figure 11 shows the jitter performance. In the figure, the proposed scheduling algorithm shows better performance than the other two schedulers. This is because our scheduler aims to empty the queue when the current transmission completes. If the queue is not empty, our scheduler begins next SI earlier to reduce the delay.

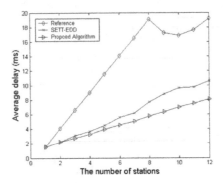

Fig. 10. Average delay in scenario 2 (VBR traffic)

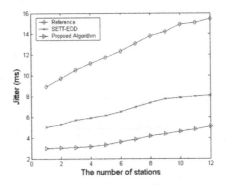

Fig. 11. Jitter in scenario 2 (VBR traffic)

5   Conclusion

Since the existing scheduling algorithms for IEEE 802.11e do not reflect well the characteristics of the traffic, it causes increase of mean delay, jitter and packet loss ratio especially when bursty VBR traffic is applied. In this paper, we proposed a new scheduling algorithm to improve the performance of HCCA. The proposed scheduler guarantees QoS by changing suitably service intervals and transmission opportunities. The service schedule is based not only on the TSPEC parameters of a flow but also on the traffic queueing. Simulation results show that the proposed scheduling algorithm is superior to the reference scheduler and SETT-EDD scheduler in terms of throughput, end-to-end delay and jitter. Simulation results showed that our scheduler outperformed the reference and SETT-EDD schedulers In this paper, our scheduling algorithm operated under assumption of a stable channel. However, The future circumstance of wireless network co-exists with various wireless technologies. More efficient scheduling study should be done under the circumstance of an unstable wireless channel.

References

1. IEEE Std, IEEE standard for wireless LAN medium access control (MAC) and physical (PHY) specification (1999)
2. IEEE Std 802.11e/D13.0, Draft supplement to standard for telecommunications and information exchange between systems-LAN/MAN specific requirements. Part 11: Wireless medium access control (MAC) and physical layer (PHY) specifications: Medium access control (MAC) enhancements for quality of service (QoS) (April 2005)
3. del Prado, J., Soomro, A., Shankar, S.: Normative Text for TGe Consensus Proposal. IEEE 802.11-02/612r0 (September 2002)
4. Prado, J.: Mandatory TSPEC Parameters and Reference Design of a Simple Scheduler. IEEE 802.11-02/705ar0 (November 2002)
5. del Prado, J., Soomro, A., Shankar, S.: TGe Scheduler - Minimum Performance Requirements. IEEE 802.11-02/ 709r0 (November 2002)
6. Skyrianoglou, D., Salkintzis, A.: Traffic Scheduling for Multimedia QoS over Wireless LANs. In: Proc. of IEEE ICC, May 2005, vol. 2, pp. 1266–1270 (2005)
7. Ansel, P., Ni, Q., Turletti, T.: An Efficient Scheduling Scheme for IEEE 802.11e. In: Proc. Modeling and Optimization in Mobile, Ad Hoc and Wireless Networks (2004)
8. Grilo, A., Macedo, M., Nunes, M.: A Scheduling Algorithm for QoS Support in IEEE 802.11E Networks. IEEE Wireless Commun., 36–43 (June 2003)
9. Skyrianoglou, D., Passas, N., Salkintzis, A.K.: ARROW: An Efficient Traffic Scheduling Algorithm for IEEE 802.11e HCCA. IEEE Transactions on Wireless Commun. 5(12), 3558–3567 (2006)
10. Stankovic, J., et al.: Deadline Scheduling for real-time Systems: EDF and Related Algorithms. Kluwer, Dordrecht (1998)
11. Ferrari, D., Verma, D.: A Scheme for Real-time Channel Establishment in Wide-Area Networks. IEEE JSAC 8(3), 368–379 (1990)
12. http://www.isi.edu/nsnam/ns
13. http://info.iet.unipi.it/~cng/ns2hcca/
14. http://www.tkn.tu-berlin.de/research/trace/trace.html

Recognition of Loneliness of the Elderly People in Ubiquitous Computing Environment

Yeong Hyeon Gu[1], Chull Hwan Song[1], Seong Joon Yoo[1], Dong Il Han[1], Jae Hun Choi[2], and Soo June Park[2]

[1] School of Computer Engineering, Sejong University, 98 Gunja, Gwangjin, Seoul, Korea
[2] Bioinformatics Team, ETRI, 161, Gajung, Yusong, Daejeon, Korea
sjyoo@sejong.ac.kr

Abstract. These days, aging of population is becoming serious problem more and more. Due to this aging, problems of suffering of unbearable pains of lonely single living old people with inconvenient movement in loneliness are occurring. Existing studies have focused only on the management of physical health of the aged but in this paper, on the contrary, psychological issues of the aged, in particular, lonely situation of the aged is reviewed. For this, system that can recognize lonely situation of the aged automatically and solve loneliness by monitoring behaviors of the aged is necessary. However, it is impossible to quantify lonely situation as there are lots of differences per each individual. In order to solve this problem, flexibility of system is required in adding or revising situations in which each individual feels loneliness. Therefore, in this paper, methods to ratiocinate lonely situation using rule are suggested. This method has advantage of no need to revise internal code as only rule can be modified when situation, in which the aged feels loneliness, is added or revised. In this thesis, system ratiocinating loneliness of the aged was designed and realized using RFID tag and reader and matters to be improved in the future are pointed out.

Keywords: Loneliness, RFID, Elderly People, Rule.

1 Introduction

These days, aging of population is becoming serious more and more. Because of this, a variety of social issues are generated and among them, the most representative issue is the health management of the aged. However, since most studies focus only on physical health of the aged, other factors of the aged such as emotion are not receiving attention comparatively. Cases of old people with inconvenient movement or lonely single living old people have not only physical pains but also mental pains caused by loneliness are also unbearable pains. To solve this, service, that can find out lonely situation of the aged and to alleviate loneliness of the aged, is necessary. Therefore, system, which can recognize and automatically solve lonely situation by Monitoring behaviors of the aged, is necessary.

In this study, RFID tags were attached to objects such as home appliances (refrigerator, water purifier, gas range, and etc.), furniture (chair, bed, sofa, and etc.),

T.-h. Kim et al. (Eds.): FGCN 2008 Workshops and Symposia, CCIS 28, pp. 60–72, 2009.

and other living supplies within homes and present status information of the aged was arranged to be collected through RFID reader attached to wrists of users for this purpose. Using information obtained like this, whether the aged are in lonely situation presently is ratiocinated and if they are in lonely situation, images helpful to alleviate loneliness were provided to the aged.

However, it was impossible to quantify since loneliness is feeling of people and lonely situation is different significantly per each person and also it is needed to consider habit of each person. To solve these problems, flexibility of system is required to add or modify situation in which each person feels loneliness. Therefore, in this system, method that ratiocinates lonely pattern of the aged using rule in XML form was suggested. Strong advantage of this method is that it is possible to add it to rule in XML format when additional modeling of lonely situation is realized in the future after a big Frame is made as the whole. Since only Rule is required to be added, it is not necessary to create module to detect new lonely situation every time and therefore, it is not necessary to modify internal code.

In the next section, related studies were reviewed and lonely situation was defined and was expressed as Rule in section 3. And in section 4, through Jess ratiocination engine, with the Rule defined in section 3, method of ratiocination was explained and in section 5, overall system structure was explained [1]. Finally, in section 6, direction and improvement points in the future were mentioned and conclusion was described.

2 Related Work

Behavior recognition studies can be classified into two kinds except present video sensor.

First, there is a system built using binary sensor of the existing theft prevention system or fire alarm system and as advantage of this system, it is cheap in price since it uses the existing theft prevention system and it is also possible to recognize by using near water or metal. In addition, it does not give burdens to users since users only need to wear relatively simple devices. On the contrary, as disadvantage, for homes without installation of theft prevention system, it is a very expensive method and it has limitation in recognizing delicate Activity because it is difficult to increase density of sensor due to the characteristics of binary sensors. In addition, it is relatively difficult to recognize and distinguish Objects compared to RFID technology. And as one of the biggest weaknesses is that it is very difficult to distinguish in case when there are many users.

STAR project, which is a representative study of this method, RFID tag, motion detect sensor, Break Beam sensor, contact sensor, and pressure sensor are installed in usual homes and through these, not only behaviors of users are recognized but also episode can be induced using recognized behaviors [2, 3]. However, tremendous amount of data association are generated as it has considered even environment with many users. So in order to solve this problem, accuracy has been increased using particle filter. However, as the number of users increase, accuracy is reduced drastically and it can be problems in handling on real time as collection time of parameter estimation is long.

As the next method, there is a method using RFID tag. Since price of RFID tag has been lowered a lot due to recent development and industrialization of RFID technology, this method has advantage of lower price and easy installation in general. And it is relatively easy to recognize and distinguish Object and also easy to distinguish multiple users. As disadvantage, due to characteristics of water or RFID, recognition is not established near metal or water due to frequency interruption and there can be problems of collision in case density of tag is increased. And since tag is detected using bracelet type Reader, it is inconvenient to use in places such as bed room or bath room.

Human Activity Recognition study developed by Intel is a representative study of this method. In this study, 14 behaviors to which Caregiver pay attention are recognized and unlike STAR project, probability of behavior is determined by mining web information in order to avoid labeling through manual works [4]. However, difference and habit of each user were not considered and it is impossible to predict next Activity if starting behavior of Activity is changed. In addition, since parameters of time were fixed, it can be difficult to recognize more realistic behaviors.

The two big behavior recognition studies were reviewed in previous part; all have used machine learning techniques based on probability. And for learning of machine learning, supervised learning was used but it can become problems since lots of time and cost are required for users to label actually large amounts of data one by one and also behavior pattern of each individual can be different.

In addition, reviewing recognition scope and object, STAR project is simple and sequential behavior and a combination of those behaviors and in the study of Intel; it is to recognize behaviors that general people or caretakers pay attention. Like this, most studies using binary sensor are focused on recognizing behaviors but in this study, it focuses on recognizing situations in which the aged feel loneliness instead of sequential behaviors.

Actually, it is very difficult to correctly find emotion of people like loneliness with binary sensor such as RFID. For this, it is necessary to have modeling regarding accurate behavior or emotion of people and it is field that more studies are required yet. Therefore, in this system, method to ratiocinate lonely patterns of the aged using rule in XML form was used. Advantage of this method, it is possible to add it to rule in XML form when additional modeling regarding loneliness is realized in the future after making a big Frame of the whole. Since it is not necessary to make a module to detect new lonely situation every time and it only needs to add rule. Therefore, it is not necessary to modify internal code. Also, it is easy to add new rules for difference and habit of each user and it is also relatively stronger for noises compared to the existing methods since time parameter is used.

3 Scenario for the Loneliness Inference System

The scenario for the Loneliness inference system for elderly that is suggested in this paper is as follows.

① Attach RFID tags to furniture or other objects in the user's house.
② The user should wear the RFID reader on their wrist.
③ The user should go about their routine activities.

④ The user is not moving even though the user's normal idle time in a chair has been far exceeded.

⑤ The user is wondering around instead of doing something specific.

⑥ Assume that these cases are when the user is feeling lonely.

⑦ Find images that can decrease their loneliness at specific times of the day.

⑧ Transmit these images to the ambient display.

This scenario is expressed in Fig. 1.

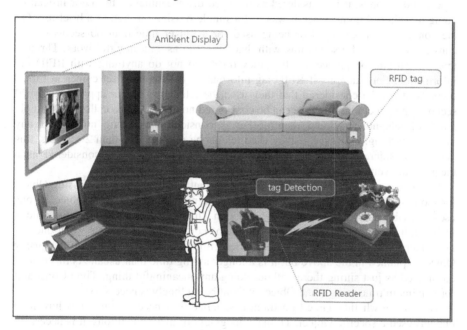

Fig. 1. Scenario Environment

RFID tags are attached to objects such as appliances (refrigerator, water purifier, stove, etc), furniture (chairs, bed, and sofa, etc) and other household goods to receive the information through the RFID reader that will be attached to the user's wrist. The purpose of the system is to interfere with the user's loneliness using the acquired information.

4 Definition of Lonely Cases

Situations to be judged as loneliness of the aged are presumed largely in two kinds. First, it is a case that the aged do not do anything for a long time and another case that the aged do only one thing for a long time on the contrary to the previous assumption. Also, the case, in which the aged do not do anything for a long time, can be classified into a case in which no RFID tag information comes in since they do not anything and another case on the contrary, in which RFID tag information comes in continuously but the aged just come and go without doing any specific thing. For this, in this study, each case is classified into three cases of 1.1, 1.2, and 2.

As it is required to ratiocinate loneliness with information and time of RFID tag collected reading through RFID Reader per each case but as it is difficult to make precise judgment with only these two kinds of information, Time Constraint was added to each object additionally. Time Constraint is a tribute to find out time using Object. If Object with Time Constraint of 60 seconds was used (contacted) in less time of 60 seconds, it is considered as the aged did meaningless behavior instead of doing meaningful things since it was just a simple passing through or a brief use. On the contrary, if it was used or being used for a longer time than 60 seconds, it is considered as they did something with that Object or as a meaningful work. Through this way, it can be judged whether they do or do not do anything with RFID tag information. For example, if RFID tag information of telephone set comes in for 20 seconds, it can be considered that they used the telephone set for approximately 20 seconds and generally, this time can be considered not as telephone calls but as simple touch of telephone set and therefore, it is considered as doing nothing or any meaningful thing. However, if they used telephone set for more than 2 minutes, this can be considered as telephone calls with somebody and it is considered as a meaningful event of telephone calls.

It is also necessary to have Time Constraint that can express facts of using Object for too long in addition to Time Constraint for minimum time of using Object. If one uses that Object for too long time even though Time Constraint for minimum time is fulfilled, it may consider it as lonely situation. For example, in a case when one sits on a chair for more than one hour, it satisfies Time Constraint of minimum time using Object explained in the above but it is a longer sitting time than usual. As this can be considered as just sitting there without doing any meaningful thing, Time Constraint for maximum time using some Object without being lonely is necessary.

To confirm whether Time Constraint is satisfied, it is necessary to know how long one has used a specific Object. To make judgment of these conditions, it is necessary to obtain Duration. Duration is classified into the following methods.

If RFID tag is sensed in RFID Reader, information of ID of RFID tag, sensed time, and user having RFID Reader are inserted into one Queue. If there is no data sensed presently, Duration is set as 0 and if there are sensed data, information of RFID tag that came into Queue presently and value of RFID tag previously came in shall be compared. If value of RFID tag detected before and value of RFID tag detected presently are not the same, it is considered as using other Object and after calculating Duration of using the Object up to now and set Duration of a new Object as 0 again. On the contrary, if values of RFID tag detected before and value of RFID tag detected presently are the same, Duration will be calculated by adding previous Duration and present Duration because the same Object is used continuously.

4.1 Case 1.1

Case 1.1 is the definition of a case in which RFID Reader does not detect any information of RFID tag because there is no movement at all. For example, let's presume a case in which the aged stands up from a sofa and moves to the center of a living room and stands there for over one hour without doing anything absent-mindedly. In this case, as there is no RFID tag that RFID Reader attached to wrist of old people can recognize within the scope in the middle of a living room, it is judged as lonely assuming that there are no activities.

Case 1.1 can be classified into two cases again and reviewing the first case, it is the case in which no RFID tag information is detected until the expiration of Threshold time after the aged first started RFID Reader. This can be more cleared if it is reviewed through Data Model. Fig. 2 expresses the data model for Case 1.1.

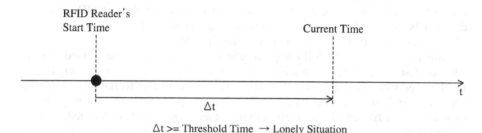

Fig. 2. Data Model for Case without any RFID Tag Information

Looking at the figure, it is noted that there was no RFID tag information sensed from the start time of RFID Reader of the aged until the current time. While renewing present time repetitively every time, Duration is obtained by deducting starting time of RFID Reader from present time. And then when value of calculated Duration becomes bigger than Threshold time, it is judged as lonely situation.

Another case of Case 1.1 is a case in which no RFID tag information is detected from the time the last RFID tag information was detected to the present. Data Model expressing such a case is like Fig. 3.

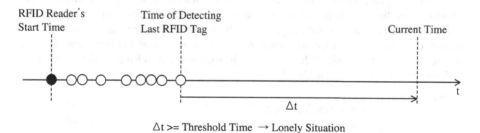

Fig. 3. Data Model for Case with More than One RFID Tag Information

Comparing the two cases, while the first case showed no detection of information of RFID tag from starting time of RFID Reader to the present, the second case is different that information of RFID tag come in continuously and from a sudden moment, information of RFID tag is not detected any more.

Even the second case, whether lonely or not is judged by comparing Threshold time and value after calculating and obtaining Duration from the time information of RFID tag is first detected to the present time just like the first case.

In this system, situation, in which there are several old people within a home, is also assumed. For example, a case, in which there are grandmother and grandfather together, can exist and to solve this case, RFID Reader with each different User ID was assigned to each old people. Information of RFID tag detected in each RFID Reader come into Sensor Data Queue and at this time, not only information of RFID tag but also information of User ID come in together as well. At this time, problematic situations occur that there is no User ID. It is noted that RFID tag is not detected because it is impossible to know User ID if RFID tag is not detected. Therefore, cases with no User Data must be handled too.

Case 1.1 consists of the following processes. First, if there is pre-defined user in advance, User ID received presently is checked. If RFID tag is detected, User ID can be identified together and if User ID is not null, this means that RFID tag is detected. Therefore, in these cases, Case 1.2 and Case 2 are inspected. On the contrary, in the case where user ID came in presently is null, as there is no RFID tag detected, present information of RFID tag and User ID are saved as none respectively after establishing present time. And for a case of zero result after checking whether User Data came in more than one time in Sensor Data Queue, Duration will be set as 0. Next, it is inspected whether User Data was none or not. If User Data came in previous time is not none, present Duration time is set as 0 and a new none information is saved after saving and renewing Duration time of using the object in the past. If User Data came in previous time is none, it was none because there was no data came in the past and as the present is the same status as it is also none, they will be integrated in the same grouping. And after calculating integrated Duration time, it is compared to Threshold time. If Duration is smaller after comparing to Threshold time, the aged is not in lonely status and on the contrary, if Duration is the same or larger than Threshold time, it is considered that they are in lonely status.

Actually, all judgments of loneliness shall be carried out in Jess ratiocination engine. However, as time concept cannot be processed in Knowledge, loneliness can be ratiocinated in Jess ratiocination engine with the information when status information is delivered to Jess after processing parts related to Duration at the early part of Jess engine. In this system, Rule of Jess was used by expressing it as Rule in XML format.

Example of the rule handling Case 1.1

```
<Rules>
    <Imp Name="Lonliness-Case1.1">
        <Head>
            <Predicate propName="contextType">
                <Variable>u</Variable>
                <Value>alone</Value>
            </Predicate>
        </Head>
        <Body operator="and">
            <Predicate propName="hasStatus">
                <Variable>u</Variable>
                <Variable>s</Variable>
            </Predicate>
```

```
                    <Predicate
    propName="durationOfNoObject">
                        <Variable>s</Variable>
                        <Variable>durNO</Variable>
                    </Predicate>
                    <FunCall funcName="greaterThan">
                        <Variable>durNO</Variable>
                        <Value>400</Value>
                    </FunCall>
                </Body>
        </Imp>
    </Rules>
```

4.2 Case 1.2

Case 1.2 is a case in which one does not do anything over a specific time while one is in motion. Reason for judging this situation as lonely status is because situations, in which one is in motion but does not do any one or more works, can be also considered as lonely.

In case when the aged are moving around a room her and there, times they use or contact each object will be all short or brief. This is considered as status, in which one does not do some works specially or focus on something, and it will be considered as lonely situation if this status continues longer than Threshold time.

As mentioned briefly in the above, we have used Time Constraint to make judgment whether one did a certain thing or not. For example, in a case when Duration time is less than 1 minute after detecting RFID tag attached to cleaner, it is difficult to consider that they did cleaning using a cleaner. With this method, objects with homes were assigned each different Time Constraint and using this Time Constraint, we can categorize roughly whether users did some meaningful works or no meaningful works.

When viewing Fig. 4, this concept can be understood more clearly. Each time when RFID tag is detected, Duration shall be compared to Time Constraint of detected object. If objects that cannot satisfy Time Constraint comes in continuously, it will be considered that the aged are in lonely situation when added time of Duration is larger than Threshold time after calculating addition of Duration of these objects.

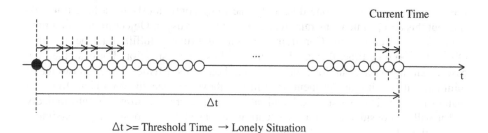

Δt >= Threshold Time $\rightarrow$ Lonely Situation

Fig. 4. Data Model for Case 1.2

After confirming first whether there are sensed data or not, duration time of sensed data will be calculated and after that, they are compared to Time Constraint of detected objects. If Duration time is larger than Time Constraint, it is considered as one did some meaningful works and on the contrary, if Duration is smaller than Time Constraint, it is considered as no meaningful works. In the next, Duration of all objects that did not satisfy Time Constraint is added all together and at this time, if addition of Duration is larger than Threshold time, it is considered as they are in lonely status.

Example of the rule for Case 1.2

```
<Rules>
    <Imp Name="Lonliness-Case1.2">
        <Head>
            <Predicate propName="contextType">
                <Variable>u</Variable>
                <Value>alone</Value>
            </Predicate>
        </Head>
        <Body operator="and">
            <Predicate propName="hasStatus">
                <Variable>u</Variable>
                <Variable>s</Variable>
            </Predicate>
            <Predicate propName="durationOfObject">
                <Variable>s</Variable>
                <Variable>durO</Variable>
            </Predicate>
            <FunCall funcName="greaterThan">
                <Variable>durO</Variable>
                <Value>400</Value>
            </FunCall>
        </Body>
    </Imp>
</Rules>
```

4.3 Case 2

Case 2 is a case in which one does only one thing continuously for a long time. To find out this case, Time Constraint that can express the use of Object for too long time is necessary. Even if Time Constraint for minimum time is fulfilled, if one uses that Object for too long time, this can be considered as lonely situation. For example, in a case of sitting on a chair for more than one hour, this satisfies Time Constraint of minimum time using object mentioned in the above but one sits on a chair too long than usual time. This is not a work and rather is considered as sitting absent-mindedly and it will be considered as lonely situation. We need Time Constraint of maximum time without being lonely while using object.

When the aged is doing one thing for too long time, it is considered as lonely and if one continues previous works after using other Object without fulfilling minimum Time Constraint momentarily while doing one thing continuously, these are no meaningful works and therefore, Duration shall be checked excluding meaningless works. If one sits on a chair continuously and use telephone for 30 seconds and again use in the middle water purifier for 45 seconds, these do not satisfy Time Constraint and as they are meaningful works, only time sitting on a chair shall be checked excluding these two events.

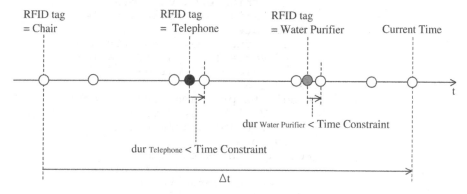

Fig. 5. Data Model for Case 2

Looking at Fig. 5, this concept can be understood more clearly. Duration is obtained after confirming whether RFID tag of the same object is detected continuously. If objects failing to satisfy Time Constraint come in the middle, addition of Duration of previous objects shall be calculated excluding these objects. When addition time of Duration is larger than Threshold time, it is considered that the aged are in lonely situation.

By confirming whether RFID tag detected presently is included in RFID tag list after checking first whether Sensor Data came in or not, if a chair or sofa is used for a long time, we check whether they are objects judged as lonely. After that, if Duration RFID tag detected presently is larger than Threshold time, it is judged as lonely situation and on the contrary, if it is smaller, all Duration of the current objects that have satisfied Time Constraint shall be added. If addition of Duration of the current objects that have satisfied Time Constraint is larger than Threshold time and if there are no other objects that also have satisfied Time Constraint in the middle, it becomes lonely situation.

Example of the rule handling Case 2

```
<Rules>
    <Imp Name="Lonliness-Case2">
        <Head>
            <Predicate propName="contextType">
                <Variable>u</Variable>
                <Value>alone</Value>
            </Predicate>
        </Head>
        <Body operator="and">
```

```
                    <Predicate propName="hasStatus">
                        <Variable>u</Variable>
                        <Variable>s</Variable>
                    </Predicate>
                    <Predicate
 propName="durationOfSameObject">
                        <Variable>s</Variable>
                        <Variable>durSO</Variable>
                    </Predicate>
                    <FunCall funcName="greaterThan">
                        <Variable>durSO</Variable>
                        <Value>400</Value>
                    </FunCall>
                    <Predicate propName="curRFObject">
                        <Variable>s</Variable>
                        <Variable>ro</Variable>
                    </Predicate>
                    <FunCall funcName="equal">
                        <Variable>ro</Variable>
                        <Value>sofa</Value>
                    </FunCall>
            </Body>
        </Imp>
    </Rules>
```

5 Inference

In this system, JESS was used for reasoning loneliness of the aged. JESS consists of Template, Rule, and Fact and it is inconvenient in modifying or updating if we use it in compliance to the grammar of JESS. Therefore, Reasoner module developed by ETRI was used for easy addition or modification of Rule whenever it is necessary. Characteristic of this module is Template and Rule written in XML form and it makes it possible to execute in JESS after parsing Fact. It is easy to modify as it only needs to modify XML file without changing internal code.

6 System Architecture

Basic function of this system is as follow. First, with information and time of RFID tag obtained through activities of the aged, loneliness of the aged is ratiocinated. Second, in case when the aged are lonely and media of images that can alleviate loneliness in compliance to the context with the results of ratiocination for the current old people shall be delivered to ambient display. Fig. 6 is a structure diagram of the system.

Role play of each module is as follow.

· SensorDataQueue: information of RFID tag read through RFID Reader, ID of RFID Reader user, and detected time of RFID tag are saved in the Queue.
· nullDetector: in case when there are no information of RFID tag read through RFID Reader, RFID tag and user ID are set as none.

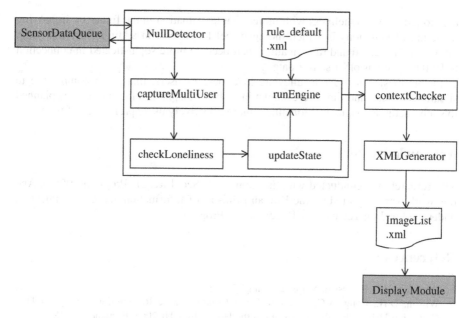

Fig. 6. Architecture of the Loneliness Detection System

· captureMultiUser: each of several users is processed respectively
· check Lonliness: check loneliness status of users with information of SensorDataQueue
· updateState: update Fact of new status in JESS ratiocination engine.
· runEngine: ratiocinate loneliness of users by executing JESS ratiocination engine.
· rule_default.xml: XML format file with the definition of Rule and Template to be used for ratiocination in JESS ratiocination engine.
· contextChecker: help users to select more proper images with seasons and time of the present.
· XMLGenerator: generate list of images in XML format to help alleviate loneliness of the aged
· ImageList.xml: image list in XML format that can help to alleviate loneliness of the aged
· Display Module: display devices that show images helpful in alleviating loneliness of the aged.

7 Conclusions

In this paper, system recommending images for alleviating loneliness in compliance of situation after recognizing loneliness of the aged with modeling loneliness status of the aged and using ratiocination technology was introduced.

There can be various situations in which the aged feel loneliness but in this system, it was judged largely by three cases of lonely situation. As this is an initial and primary demo system, the number of cases is few and simplified but it needs to be able to judge even loneliness with more complex situation in the future by improving modeling of situations in which the aged feel loneliness. In addition, by improving ratiocination method and ratiocination shall become more sophisticated in connection to RFID as well as other sensor data.

In this thesis, in order to alleviate loneliness of users, images in compliance to situations of date and time were shown in ambient display. In the future, it is planned to study methods to alleviate loneliness more effectively using preference of users.

Acknowledgement

This research was conducted with the support of 'Seoul R&BD Program(10581). And this work was supported by the Korean Ministry of Information and Communication under the Leading Technology Development Program.

References

1. Friedman-Hill, E.: Jess in action. Manning (2004)
2. Wilson, D.H., Long, A.C., Atkeson, C.: A Context-Aware Recognition Survey for Data Collection Using Ubiquitous Sensors in the Home. In: CHI 2005, Portland, pp. 1865–1868 (2005)
3. Wilson, D.H., Atkeson, C.: Simultaneous Tracking and Activity Recognition (STAR) Using Many Anonymous, Binary Sensors. In: Gellersen, H.-W., Want, R., Schmidt, A. (eds.) PERVASIVE 2005. LNCS, vol. 3468, pp. 62–79. Springer, Heidelberg (2005)
4. Philipose, M., Fishkin, K.P., Perkowitz, M., Patterson, D.J., Fox, D., Kautz, H., Hahnel, D.: Inferring Activities from Interactions with Objects. IEEE Pervasive Computing 3(4), 50–57 (2004)

Voronoi-assisted Parallel Bidirectional Heuristic Search

Anestis A. Toptsis and Rahul A. Chaturvedi

Dept. of Computer Science and Engineering,
York University, Toronto, Ontario, Canada
{anestis,rahul}@cse.yorku.ca

Abstract. We propose a method for identifying well-placed island nodes for the purpose of performing a bidirectional parallel heuristic search algorithm. Multi-process bidirectional heuristic search algorithms that utilize island nodes (such as PBA*) have been shown to have the potential for exponential speedup over their plain counterparts that do not utilize island nodes. The problem of how to generate appropriately located island nodes has resisted any general purpose solution to date. The proposed method is an initial step toward this end. We implement our method and evaluate its performance within PBA* for a variety of sliding-tiles puzzles. Our findings reveal that the overhead cost of using our method is negligible, while at the same time, when PBA* is equipped with the proposed method, it outperforms its random-island-nodes counterpart for the vast majority of test cases.

1 Introduction

Heuristic search is one of the foundational areas of artificial intelligence (AI). Recently, it has many applications in several diverse and practical areas such as Software Engineering and the Web (e.g., [1, 2, and 3]). Heuristic search is essentially a graph traversal with the characteristic that the number of nodes in the graph is *huge* and thus the graph nodes are generated on the fly (as opposed to being already known), as we traverse the graph. This sort of traversal is called heuristic search, because we always search for some "goal" node G, in the graph, and we always use heuristics (guesses and estimates that indicate the best way to proceed in the course of the graph traversal). Heuristic search algorithms can be classified into two types, *unidirectional* and *bidirectional*. In the unidirectional category (e.g., [4]) there is only one-direction type of process, emanating from the source node S and seeking the goal node G. The bidirectional category (e.g., [5, 6, 7, and 8]) incorporates two types of processes – one forward type search process, from S to G, and one reverse type search process, from G to S. Each of the processes can be executed using a typical unidirectional algorithm such as A* [4]. In some bidirectional search algorithms, special kind of nodes, called X-nodes (or islands), are used to aid the search. The concept of introducing X-nodes to speed up the search was introduced in [9] and there have since been developed several algorithms that utilize X-nodes. Notable examples are PBA*, WS_PBA*, and SSC_PBA*, described in [7, 8].

Bidirectional multiprocessor heuristic search has been introduced by Nelson [13], first for 2 processors as a parallelization of Pohl's algorithm [5], and it was later generalized

T.-h. Kim et al. (Eds.): FGCN 2008 Workshops and Symposia, CCIS 28, pp. 73–84, 2009.
© Springer-Verlag Berlin Heidelberg 2009

for N processors, N > 2 [11, 12]. The N-processor algorithm, PBA*, uses intermediate nodes of the search space, called *islands* or *X-nodes*. The terms *islands* and *X-nodes* are equivalent for the purposes of this paper. Historically, the term *island* nodes first appeared in [9] to denote intermediate nodes in the search space, with the property that an optimal path passes through at least one of them. The term *X-nodes* was introduced in [11] to denote intermediate node in the search space but not requiring that an optimal path passes through at least one of them. Also, since the algorithm described in [11] requires two search processes (one forward and one reverse) to emanate from each node, the term *X-node* was coined as a reminder of the bidirectionalism in the search (see figure 1) in order to divide the search space into smaller subspaces.

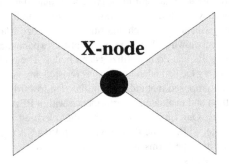

Fig. 1. X-node

In this paper, the terms *islands* and *X-nodes* are used interchangeably, and mean intermediate nodes of the search space, some of which may participate in a path (not necessarily optimal) connecting the source and goal nodes. Note, some of these intermediate nodes may not participate in a solution (actually this is especially true in [9] where only one of the island nodes participates in the solution). As illustrated in Figure 2, in addition to the parallel searches conducted from the source node S and the goal node G, two parallel searches are conducted from each X-node; a *forward* search towards the goal node G, and a *reverse* search towards the source node S. A solution is found as soon as a set of search spaces intersect in such a way that a path can be traced from S to G. The complexity of PBA* was analyzed in [11, 12]. In the

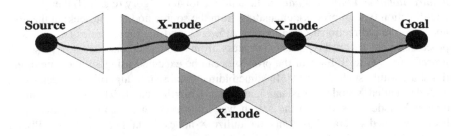

Fig. 2. Parallel bidirectional search with islands

case that the X-nodes are placed equidistantly on an optimal path from source to goal, the complexity of PBA* is

$$O\left(b^{\frac{n}{2\cdot(|X|+1)}}\right)$$

where b is the branching factor of the problem (branching factor is the number of nodes generated at any node expansion), n is the length of an optimal path from source to goal, and |X| is the number of X-nodes (or islands). This situation is illustrated in Figure 3 for |X| = 2.

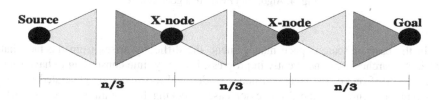

Fig. 3. Algorithm PBA* in the best case

It can be shown (although the proof is not given in this paper) that the speedup of PBA* over its uniprocessor version is potentially *linear on the number of processors*, and the speedup of PBA* over the conventional uniprocessor A* algorithm is potentially *exponential on the number of processors*. By "uniprocessor version of PBA*" it is meant that algorithm PBA* is simulated on a uniprocessor machine. This can be done, for example, by using a simulating program of a parallel machine running on a uniprocessor (such programs may be provided by the manufacturer of the parallel computer and are usually used for software development prior to porting to the actual parallel machine), or by simply spawning processes in a multitasking operating system like UNIX, or by utilizing the multithreading facility of modern programming languages and operating systems.

In the general case where X-nodes are not placed equidistantly on an optimal path from source to goal, the complexity of PBA* is

$$O\left(b^{\frac{\max\left(dist\left(X_i,X_j\right),dist\left(X_i,E\right)\right)}{2}}\right)$$

where X_i and X_j range over all X-nodes, and E is either the source or the goal node. This situation is illustrated in Figure 4 for two X-nodes. There are two main issues regarding the efficiency of PBA*. The first is search space management, that is, how to efficiently detect intersections among the many search spaces. The second is the X-node generation, that is, how to find X-nodes on a path from source to goal.

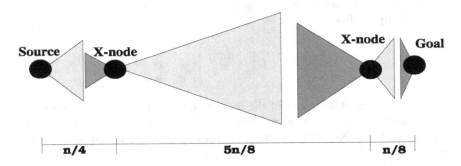

Fig. 4. Algorithm PBA* in a general case

In the case of search space management, the difficulty arises from the fact that when N searches are concurrently in progress, it is very time consuming to have each process test if one of its generated nodes is identical to some node generated by any other process (note, in the 2-processor case, checking for common nodes is fairly inexpensive, since there are only two searches in progress). As an attempt to ease this task, *multi-dimensional heuristics* (MDH) were developed by [11, 12]. As part of the MDH approach, additional nodes, called *reference* nodes, are designated into the search space, in order to "guide" the search. Based on the reference nodes use in MDH, several *intersection detection algorithms* have been devised [11, 10, and 12]. The algorithms' task is to prevent node exchange between any two processes, unless there is a high probability that the search spaces of these processes intersect. Simulation results have shown that in the worst case, unnecessary node exchange is cut down in half, whereas in the best case it is cut down almost completely (only 1% of the nodes are exchanged between two non-intersecting subspaces). Later in this paper we outline some of the intersection detection algorithms since we use them in the implementation of our proposed method here.

2 Our Proposal

We present our method of how to select X-nodes which are located appropriately in the state space. Algorithm A, below, describes the proposed method.

Algorithm A.

Step A.1: generate a set $X = \{x_1, ..., x_M\}$ of *candidate* X-nodes.

Step A.2: convert set X into set $P = \{\pi_1, ..., \pi_M\}$ such that π_i is the projection of x_i onto a 2-dim space.

Step A.3: Form D-graph, the Delaunay graph for set P.

Step A.4: Calculate the *shortest* path SP = [S-X1-X2- ...-Xk-G] connecting S and G in the D-graph of step A.3.

Step A.5: Select as X-nodes the nodes corresponding to points X1, X2, ..., Xk and execute algorithm PBA* between S and G.

In *step A.1*, we randomly generate a fairly large number of nodes. These are *candidate* X-nodes and the intention is that after the processing done in steps A.2, A.3, and A.4, a small number of those nodes is selected for being used in algorithm PBA*. In *step A.2*, we map the nodes x_i generated in step A.1 onto points π_i of an N-dim space. In our case, we use N=2. (This is because a 2-dim space is more convenient to use for step A.3. In the future we plan to extend our work for N > 2). To project a node x_i onto a point π_i in 2-dim space we use two randomly generated nodes r_1 and r_2 (other than the ones of set X of step A.1). Then π_i is calculated as

$$\pi_i = \left(md\left(x_i, r_1 \right), md\left(x_i, r_2 \right) \right)$$

where $md\left(x_i, r_j \right)$ is the Manhattan distance between x_i and r_j (j=1, 2). In *step A.3*, we form the D-graph of the set P calculated in step A.2. This graph is the Delaunay triangulation DT(P) for the set of points P. The main property of DT(P) is that no point in P is inside the circumcircle of any triangle of DT(P). Also, the DT(P) corresponds to the dual graph of the Voronoi diagram for P. This latter property is our main motive in calculating the DT(P). Without loss of generality, we assume that the vertices of DT(P) correspond to the centers/generators of the Voronoi cells in the Voronoi diagram that corresponds to DT(P). The property of interest from any Voronoi diagram is that all the points within a Voronoi cell are closer to the center of that cell than to the center of any other cell, including the neighboring cells. Therefore, by considering the Voronoi diagram of the DT(P) graph, we somewhat avoid choosing as X-nodes nodes that are too close to each other. (Note, such X-nodes would correspond to search spaces that possibly duplicate each other and thus waste resources by exploring the same part of the global search space). Consequently, we designate as our X-nodes the generator nodes of the Voronoi cells that form a *shortest path* between S and G. In *step A.4*, we apply Dijkstra's shortest path algorithm on graph DT(P) and calculate SP = [S-X1-X2- ...-Xk-G], the shortest path connecting S and G. Note, S and G of path SP of step A.4 are not the actual source and goal *nodes* but they are the 2-dim *points* (as calculated in step A.2) corresponding to the actual source and goal nodes. Also, X1, ..., Xk of step A.4 are 2-dim points corresponding to certain candidate X-nodes generated during step A.1. For simplicity in notation we use the same symbols to refer to the corresponding actual nodes S and G and the corresponding X-nodes of the 2-dim points X1, ..., Xk. In *step A.5*, we designate as our X-nodes the nodes that correspond to the points X1, ..., Xk, and using those nodes we execute algorithm PBA* between S and G.

3 Performance Evaluation

We implement A*, the traditional heuristic search algorithm, and two versions of PBA*, PBA*-R and PBA*-VD. PBA*-R is PBA* with randomly generated X-nodes; PBA*-VD is PBA* with Voronoi-Dijkstra designated X-nodes, per Algorithm A of Section 2. We use the sliding tiles puzzle problem for our tests and run A*, PBA*-R, and PBA*-VD for 17 puzzles (3 of the puzzles are 6x6, i.e., 35-puzzles, 4 are 5x5, and 10 are 4x4 puzzles). We use the Manhattan distance as our heuristic function. Algorithms PBA*-R and PBA*-VD complete their execution and find solutions for all 17 puzzles. Due to memory space limitations, algorithm A* is not able to find a solution for any of the 6x6 puzzles, and for 2 of the 5x5 puzzles. Although algorithms PBA*-R and PBA*-VD are not admissible, they find near-optimal solutions. (In fact, the average solution path length found by both versions of PBA* is near-equal to the optimal path length, 45, as found by A*, which is an admissible algorithm.). For either version of PBA*, an *intersection detection algorithm*, IDA-3, is used to control search process communication. IDA-3 is originally described in [10] and it has been used in several of our previous works (e.g., [7, 8, 14, 15, 16]). The main characteristic of IDA-3 is that it instructs two opposite-direction search processes to exchange nodes (and, henceforth, compare those nodes) if such nodes are deemed to be "similar enough" so that they are possibly identical. Since algorithm IDA-3 is central in our evaluation due to its communication cost, we describe IDA-3 here.

Intersection Detection Algorithm IDA-3.

Every search space is represented as a N-dimensional polyhedron with 2^N corners. Each corner has coordinates

$$\left(R_{1i}, R_{2i}, ..., R_{Ni}\right)$$

where R_{ji} is either the minimum or the maximum distance of the X-node corresponding to that search space, from the reference node R_j, j = 1, ..., N. Two search spaces S_a and S_b may contain a common node when there is an overlap of their corresponding polyhedra. A common node between two search spaces is not possible to exist, until such an overlap occurs.

Two search spaces S_a and S_b might contain an intersection when there is an overlap for each of their reference node ranges. Pictorially this is represented by an intersection of the approximated spaces in the N-space. In Figure 5 this occurs for S_3 and S_4 (for 2 reference nodes).

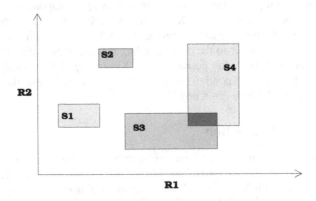

Fig. 5. N-space intersection in PBA* (N = 2)

In algorithm PBA* a *central control process* (CCP) is used to coordinate the local search processes running concurrently. The CCP compares the search spaces with respect to their estimated distances to the reference nodes and instructs two search processes to start exchanging nodes when their search spaces seem to be intersecting. As it is quite expensive to store *all* the reference nodes' distances for *all* nodes generated by each search space, the CCP uses an *overlap table* to store selected distances for each reference node. Specifically, the overlap table holds only the minimum and maximum distances to each reference node from each search space. Each search space keeps track of its minimum and maximum values for each reference node, and informs the CCP as these values change. The CCP receives the new values, updates the overlap table, and checks if an overlap between a pair of opposite direction search spaces has occurred. An overlap with respect to a reference node k occurs when

$$\left(S_i R_k \min, S_i R_k \max \right) \cap \left(S_j R_k \min, S_j R_k \max \right) \neq \varnothing$$

$$S_j R_k \min = \min \left\{ dist \left(n_{S_j}, R_k \right) \right\}$$

and

$$S_j R_k \max = \max \left\{ dist \left(n_{S_j}, R_k \right) \right\}$$

for any node n_{S_j} in S_j.

It happens that an intersection between two search spaces is not possible until such an overlap occurs. Figure 6 is an example of an overlap table maintained by the central control process when four search processes and four reference nodes are present. The table shows that search processes S_1 (forward) and S_2 (reverse) overlap with respect to all four reference nodes. In this case, the CCP will instruct search processes S_1 and S_2 to start node exchange. In particular, the reverse search process S_2 is instructed to send nodes to the forward search process S_1, and the forward search process S_1 is given an alert that nodes from S_2 are to arrive.

Search process	R1		R2		R3		R4	
	Min	Max	Min	Max	Min	Max	Min	Max
S1 (forward)	10	17	63	68	4	25	9	26
S2 (reverse)	14	27	43	65	13	31	18	40
S3 (forward)	4	7	11	19	4	12	4	19
S4 (reverse)	28	32	22	31	37	51	28	43

Fig. 6. A sample overlap table

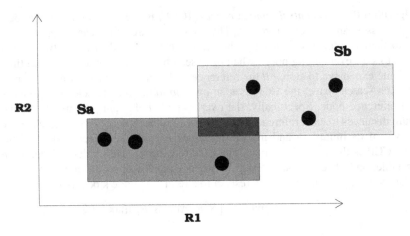

Fig. 7. S_a and S_b overlap, but $S_a \cap S_b = \varnothing$

If the central control process uses an overlap table to decide when to turn on communication between two searches, it can be assured that some existing intersection will not be overlooked. Unfortunately, the existence of these overlaps does not insure that the two search spaces contain a common node. The condition is not sufficient for several reasons, one of which is that the heuristics only *estimate* distances to the reference nodes. However, even if the heuristic makes no errors in estimating distances, there are still other reasons why this condition is not sufficient. As shown in Figure 7, it may be that the approximated space of S_a intersects with the approximated space of S_b without having any single node in (S_b) S_a overlap all the reference node ranges in (S_a) S_b. The result of this is that no nodes lie in the intersection of the approximated spaces [11]. Three different intersection detection algorithms, which increasingly progressive degree of refinement, are outlined next.

(1) *IDA-1* maintains a [min, max] overlap table and when the approximated spaces of S_f and S_r intersect, S_r will begin sending its nodes to S_f.

(2) *IDA-2* is based on addressing one of the reasons which keeps the overlap condition, the basis of IDA-1, from being sufficient to insure that two spaces intersect. IDA-2 waits until the reference node values of a single node in S_r lie in the approximated space of S_f. When this condition occurs, S_r will then begin sending its nodes to S_f.

(3) *IDA-3* checks for the same condition as IDA-2 except that it does so on a node by node basis. Both IDA-1 and IDA-2 end up turning on communication permanently. It may be that two search spaces only come close to intersecting for a short period of time resulting to a lot of unnecessary node sending if communication is permanently turned on. Instead, IDA-3 checks as each node is generated and only sends nodes which lie in the approximated space of S_f.

As mentioned earlier, we adopt IDA-3 for our evaluation. Our experiments reveal the following.

Result 1: For the vast majority of tests (more than 80% of test cases), the X-nodes used in algorithm PBA*-VD are more likely to aid in establishing a path connecting S and G than the X-nodes used by algorithm PBA*-R. This is illustrated by Table 1.

Table 1. IDA-3 probes for possible search space intersection

	PBA*-R	PBA*-VD	Winner
	16,642	25,937	**PBA*-VD**
	15,573	38,015	**PBA*-VD**
	27,980	27,462	**PBA*-R**
	35,753	43,784	**PBA*-VD**
	1,106	3,380	**PBA*-VD**
	28,756	29,584	**PBA*-VD**
	56,574	72,625	**PBA*-VD**
	15,605	19,599	**PBA*-VD**
	239,480	231,296	**PBA*-R**
	61,741	73,515	**PBA*-VD**
	28,598	31,534	**PBA*-VD**
	227,819	235,092	**PBA*-VD**
	150,297	153,509	**PBA*-VD**
	48,063	58,671	**PBA*-VD**
	165,146	205,781	**PBA*-VD**
	69,090	85,686	**PBA*-VD**
	143,781	132,448	**PBA*-R**
Total	1,332,004	1,467,918	
Average	78,353	86,348	
wins of PBA*-VD over PBA *-R			**82.35%**

Note, for all but three cases in Table 1, algorithm PBA*-VD probes more times for search space intersection.

Result 2: The overhead for incorporating the Voronoi-Dijkstra method in finding useful X-nodes is *negligible*. This is illustrated by Table 2.

Table 2. Total overhead (over all test puzzles)

method	Total (sec)
PBA*-R	**6,435**
PBA*-VD	**6,103**
GRN overhead	**16.437**
D/D overhead	**8.819**
Total Overhead (GRN + D/D)	**25**

As shown in Table 2, there are two types of overhead in incorporating Algorithm A of section 2 into algorithm PBA*: the *GRN overhead*, and the *D/D overhead*. The GRN overhead is the cost of executing essentially steps A.1 and A.2 of Algorithm A (generate candidate X-nodes and project them onto a 2-dim space). The D/D overhead is the cost of executing steps A.3 and A.4 of Algorithm A (form the D-graph and calculate shortest path with Dijkstra's algorithm). As we see in Table 2, the *total overhead* (GRN + D/D) is 25 seconds, which is 0.41% (25/6103) of the time required to execute algorithm PBA*-VD. We also note that the total time to execute PBA*-VD (i.e. the time for the actual PBA*-VD plus the total overhead time) does not exceed the time to execute PBA*-R. Therefore, assuming that the incorporation of the Voronoi-Dijkstra technique into algorithm PBA* does not, in any way, harm the overall quality of the heuristic search process, the overhead for employing the Voronoi-Dijkstra method for finding useful X-nodes is not only negligible, but it also positively contributes (as evidenced by the results shown in Table 1), to the overall quality of PBA*.

Result 3: *The Voronoi-Dijkstra "anomaly"*. Our experiments uncover an unfortunate scenario, illustrated in Figure 8.

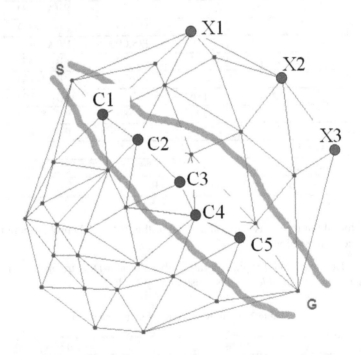

Fig. 8. The Voronoi-Dijkstra "anomaly"

Figure 8 shows the D-graph formed by a specific arrangement of randomly generated candidate X-nodes. The path S, X1, X2, X3, G is the shortest path connecting S and G and, therefore, X1, X2, and X3 are chosen as the X-nodes for the execution of algorithm PBA*-VD. Note, however, although these nodes are the ones that form a shortest path between S and G, nodes C1, ..., C5 seem to be a better alternative for X-nodes to use for PBA*-VD! This leads us to speculate that the *shortest path* (as calculated by the Dijkstra algorithm) *may not be the best choice* for X-nodes and, instead, the *straightest path* between S and G might be a *better alternative*! Investigation of the ramifications of this "anomaly" is in our immediate research plans.

4 Conclusion

We present a method to generate appropriately located island nodes within a search space. The motive for doing so is that such generated nodes will help in establishing a solution path faster, if used by a multi-process bidirectional heuristic search algorithm, such as PBA*. To the best of our knowledge this problem has resisted any type of general purpose solution for more than two decades. We implement our method and test it using PBA*, a bidirectional multi-process heuristic search algorithm designed to utilize island nodes. Our findings indicate that PBA*-VD (the version of PBA* that uses the island nodes generated by our method) outperforms PBA*-R (the version of PBA* that uses randomly generated island nodes), more than 80% of the time. Also, the overhead of incorporating our method into PBA* is negligible (less than 0.5% of the cost of executing the PBA* algorithm itself). Interestingly, we also uncover an "anomaly" (Result 3, in Section 3), whose remedying points to a method of generating even more appropriately located island nodes.

Our future research plans include investigating the ramifications of the found "anomaly" and also extending our method to N-dimensional spaces, when N > 2. Also, we are interested in investigating the impact of the placement of reference nodes (the use of reference nodes is at the heart of the MDH methodology described in section 2). Specifically, besides the placement of X-nodes, the placement of reference nodes possibly impacts the performance of PBA*. In our current implementation of PBA*-R and PBA*-VD we use randomly generated reference nodes – which is the approach followed by all previous works in this area. Earlier findings (e.g., [10]) suggest that increasing the number of reference nodes beyond a certain threshold (about five reference nodes) provides little or no improvement on the performance of PBA*. However, the impact of the location of reference nodes is, as yet, unknown. We are currently working toward investigating this problem.

References

1. Antoniol, G., Di Penta, M., Harman, M.: Search-Based Techniques Applied to Optimization of Project Planning for a Massive Maintenance Project. In: Proceedings of IEEE International Conference on Software Maintenance, pp. 240–249 (2005)
2. Yu, F., Ip, H.S., Leung, C.H.: A Heuristic Search for Relevant Images on the Web. In: Leow, W.-K., Lew, M., Chua, T.-S., Ma, W.-Y., Chaisorn, L., Bakker, E.M. (eds.) CIVR 2005. LNCS, vol. 3568, pp. 599–608. Springer, Heidelberg (2005)
3. Bekkerman, R., Zilberstein, S., Allan, J.: Web Page Clustering using Heuristic Search in the Web Graph. In: Proceedings of IJCAI 2007, the 20th International Joint Conference on Artificial Intelligence (2007)

4. Hart, P.E., Nilsson, N.J., Raphael, B.: A Formal Basis for the Heuristic Determination of Minimum Cost Paths. IEEE Transactions on Systems, Science, and Cybernetics 4(2), 100–107 (1968)
5. Pohl, I.: Bi-Directional Search. Machine Intelligence, 127–140 (1971)
6. DeChampeaux, D.: Bidirectional Heuristic Search Again. Journal of the ACM 30(1), 22–32 (1983)
7. Nelson, P.C., Toptsis, A.A.: Superlinear Speedup Using Bidirectionalism and Islands. In: Proc. International Joint Conference on Artificial Intelligence (IJCAI) - Workshop on Parallel Processing in AI, Sydney, Australia, pp. 129–134 (1991)
8. Nelson, P.C., Toptsis, A.A.: Unidirectional and Bidirectional Search Algorithms. IEEE Software 9(2), 77–83 (1992)
9. Chakrabarti, P.P., Ghose, S., Desarkar, S.C.: Heuristic Search Through Islands. Artificial Intelligence 29, 339–348 (1986)
10. Nelson, P.C., Henschen, L.: Multi-Dimensional Heuristic Searching. In: IJCAI 1989 - International Joint Conf. on Artificial Intelligence, pp. 316–321 (1989)
11. Nelson, P.C.: Parallel Bidirectional Search Using Multi - Dimensional Heuristics, Ph.D. Dissertation, Northwestern University, Evanston, Illinois (June 1998)
12. Nelson, P.C.: Parallel Heuristic Search Using Islands. In: Proc. 4th Conf. on Hypercubes, Concurrent Computers and Applications, Monterey (March 1989)
13. Nelson, P.C., Henschen, L.: Parallel Bidirectional Heuristic Searching. In: Proc. Canadian Information Processing Society, Montreal, Canada, vol. 5, pp. 117–124 (1987)
14. Toptsis, A.A., Nelson, P.C.: Parallel Bidirectional Heuristic State-Space Search. Heuristics Journal 6(4), 40–49 (Winter 1993)
15. Toptsis, A.A.: Parallel Bidirectional Heuristic Search with Dynamic Process Re-Direction. In: Proc. 8th International Parallel Processing Symposium, IPPS 1994, pp. 242–247. IEEE Computer Society Press, Los Alamitos (1994)
16. Nelson, P.C., Toptsis, A.A.: Search Space Clustering in Parallel Bidirectional Heuristic Search. In: Proc. 4th UNB Artificial Intelligence Symposium, New Brunswick, Canada, pp. 563–573 (September 1991)

Artificial K-Lines and Applications

Anestis A. Toptsis and Alexander Dubitski

Dept. of Computer Science and Engineering,
York University, Toronto, Ontario, Canada
{anestis,dubitski}@yorku.ca

Abstract. We propose Artificial K-lines (AKL), a structure that can be used to capture knowledge through events associated by causality. Like Artificial Neural Networks (ANN), AKL facilitates learning by capturing knowledge based on training. Unlike and perhaps complimentary to ANN, AKL can combine knowledge from different domains and also it does not require that the entire knowledge base is available prior to the AKL usage. We present AKL and illustrate its workings for applications through two examples. The first example demonstrates that our structure can generate a solution where most other known technologies are either incapable of, or very complicated in doing so. The second example illustrates a novel, human-like, way of machine learning.

1 Introduction

Artificial Intelligence (AI) is enjoying a renewed interest which makes its presence welcome in many aspects of our daily life. Even before, and certainly since the appearance of the phrase "AI", the following questions are of utmost importance: How come people are able to *learn so much*? How come people *are creative* (i.e., able to perform a new task, different from two or more previously learned tasks, by being "inspired" by their previous experiences and knowledge)? These issues have puzzled philosophers and cognitive scientists for many years, and with the appearance of AI as a field related to those disciplines, they are among the core AI questions as well. Numerous attempts to provide an overall answer to these issues, failed during the past 50 years. However, several "theories of memory" have emerged. Notable example are [1], [2], [3], and [4]. None of these approaches has been fully implemented to date; however, there have been several reports toward this end. Examples are [5] and [6]. A central theme in [1], [2], and [3] is the concept of K-lines. Quoting from [1],

> When you "get an idea," or "solve a problem" [...] you create what we shall call a K-line. [...]...When that K-line is later "activated", it reactivates [...] mental agencies, creating a partial mental state "resembling the original".

In some of our previous works (e.g. [7]) we have used a form of K-lines to address media handling issues in affective computing systems. In this paper, inspired by the concept of K-lines, we introduce a structure that can possibly exhibit the caliber of intelligence usually attributed to Artificial Neural Networks (ANN). To the best of our knowledge, no such proposal exists for using K-lines in the way described here. The rest of the paper is organized as follows.

T.-h. Kim et al. (Eds.): FGCN 2008 Workshops and Symposia, CCIS 28, pp. 85–97, 2009.
© Springer-Verlag Berlin Heidelberg 2009

Section 2 describes the proposed structure. Section 3 provides, through two examples, a discussion of two possible applications of AKL. Section 4 gives a comparison of our method with the essentials of the ANN structure. Section 5 is the conclusion and ideas for future research.

2 Our Proposal

We define a K-line to be a *sequence of associated events* $e_1, e_2, ..., e_k$, such that e_i and e_{i+1} are connected by *causality*. That is, the occurrence of event e_{i+1} is a direct consequence of the occurrence of event e_i. If two K-lines contain the same event, then they are said to *intersect* at that event. When two K-lines are created, if they do not intersect, they form a graph like the one shown in Figure 1. If the K-lines intersect, then they form a graph like the one shown in Figure 2.

In Figure 1, the nodes (circles in the graph) of each K-line represent events. An edge such as (e11, e12) means that event e12 is a consequence of event e11. In figure 1, K-lines KL1 and KL2 do *not* intersect. Consequently, the graph provides two possible "ways to think", as shown in parts (a) and (b) of Figure 1.

In Figure 2, K-lines KL1 and KL2 intersect. Consequently, the graph provides four possible "ways to think", as shown in Figures 3 (a), (b), (c), and (d).

Each way of thinking is created by starting from one of the available K-lines, and then, once we encounter an intersection, we follow each of the two possible paths. Note, for every intersection that we encounter, the number of "ways to think" is multiplied by the outdegree of the node that is at that intersection. Therefore, the more intersections we have, the more times the number of "ways to think" is multiplied. This is the crucial point in our method.

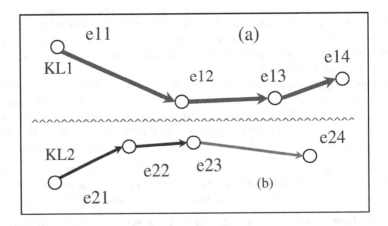

Fig. 1. Two K-lines that do not intersect

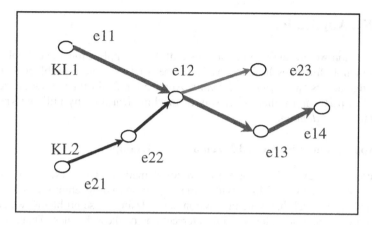

Fig. 2. Two intersecting K-lines

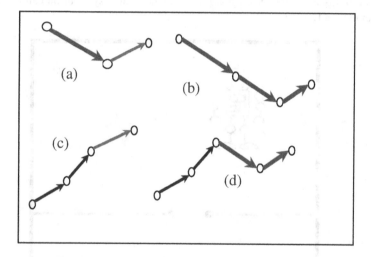

Fig. 3. Four possible ways to think: (a), (b), (c), (d)

Note, by combining 2 K-lines (KL1 and KL2) we have a system (Fig. 2) capable to "think" in *more* ways than is possible to think if the 2 K-lines are not combined (Fig. 1). By adding more intersecting K-lines, it is expected that the number of "ways to think" is increased *much faster* than the number of the added K-lines. Also note, some of the possible formed "ways to think" are comprised by *parts* of existing K-lines. This is the essence of creativity, i.e., to generate a new idea by using parts of old ideas. The above discussion certainly does not constitute proof, it, nevertheless, fits the essence of how people are able to learn *so much* (by forming intersecting K-lines in their brains) and also how people are *capable of creativity* (by forming new ideas from parts of some of the formed K-lines).

3 AKL Applications

In this section we present two possible uses of AKL, Application 1 and Application 2. Application 1 illustrates how the AKL can be used to solve a problem by utilizing knowledge across two *different domains*. Application 2 illustrates how the AKL can be used to perform a rather novel way of machine learning, by utilizing knowledge within the *same domain*.

3.1 Application 1: Robot and Assembly Line Example

We describe an example of how the proposed method can work, and illustrate its benefits. We create two K-lines with events based on two different scenarios that can occur in hypothetical, but realistic environments. Using those problems, we create K-lines of events and note the possible intersections of these K-lines. Then, we pose a task to be performed and we show that the AKL graph is capable to easily produce a solution to this task, whereas it is not obvious how an easy solution can be produced without the AKL graph.

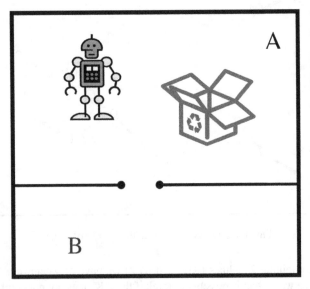

Fig. 4.a. (a) Robot has to move box from area A to area B through opening. (b) Assembly line for processing boxes.

The first problem is a robot movement problem. This problem is described in [5] and it is used here to apply our method for its solution. The second problem is a situation that can occur as part of the operation of a typical assembly line. Figure 4(a) illustrates the situation for the first problem and Figure 4(b) illustrates the situation for the second problem. In Figure 4(a), the robot is needed to move the box, from room A to room B, passing it through the opening that connects the two rooms. In Figure 4(b), a box is on a moving belt. As part of the assembly line process, the first worker unfolds

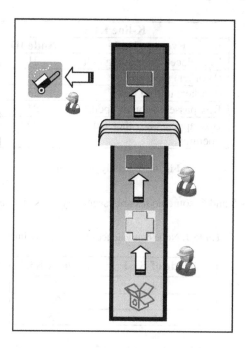

Fig. 4.b. Assembly line for processing boxes

the box and the second worker flattens the box further and places it on the moving belt again. Then, the moving belt carries the box through an opening and the flattened box reappears at the other side of the opening. After that, further processing of the box may occur, but this is not relevant to our example.

A K-line of events that captures a possible process of the robot moving the box through the opening is shown in Figure 5.

K-line KL1	
Event in Node	**Node ID**
(robot) Move toward box	11
(Robot) pick up box	12
Move box toward opening	13
Move box through opening	14
Box passed through opening	15
Box appears at other side of opening	16

Fig. 5. Robot K-line

A possible K-line of events that captures the process of the assembly line is shown in Figure 6.

K-line KL2	
Event in Node	**Node ID**
Box placed on moving belt	21
(Worker) pick up box	22
(worker) unfold box	23
Box passed through opening	24
Box appears at other side of opening	25

Fig. 6. Assembly line K-line

Note, in Figures 5 and 6, some nodes are essentially the same, as shown in Table 1.

Table 1. Node equivalences between 2 K-lines

Node of KL1	**Node of KL2**
12	22
15	24
16	25

Each row of Table 1 shows two nodes, one from K-line KL1 and one from K-line KL2. Each row of Table 1 contains nodes that are the same, or essentially equivalent, according to the K-lines of Figures 5 and 6. Based on this observation, the two separate K-lines are said to *intersect*, and they can be merged into a graph, as shown in Figure 7.

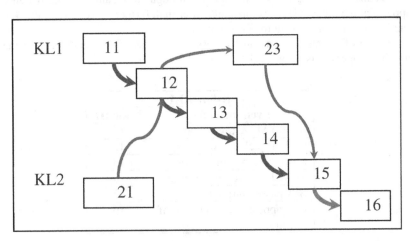

Fig. 7. Intersecting K-lines

Note, in Figure 7, the edge connecting nodes 15 and 16 belongs to both K-lines. We argue that *the knowledge conveyed by Figure 7 is superior to the knowledge conveyed individually by Figures 5 and 6*. Next, we explain why this is the case.

Consider the following problem: In a situation like the one described in Figure 5, we would like the robot to move the box from room A to room B. Also, we assume that the box is too big to fit through the opening that connects the two rooms, but the robot doesn't know this! (The same problem with the same assumption is considered in [5], where a method based on reflective planning is developed to solve it). If we use the *individual* K-lines of Figure 5 or Figure 6 (i.e., without those K-lines being merged), then there are two possible solutions, S1 and S2: (S1) start from K-line KL1 and produce the sequence 11-12-13-14-15-16; (S2) start from K-line KL2 and produce the sequence 21-22-23-24-25. Note, *S1* will *fail* at step 14 since the box does not fit through the opening between rooms A and B; and *S2* is *not applicable*, since the starting point, 21, is outside the domain of the robot problem! Therefore, the system of two *separate* K-lines KL1 and KL2 is incapable of solving this problem.

However, if we consider the same K-lines but with the K-lines being merged, (as shown in Figure 7), then we can generate 4 (instead of 2) possible solutions: (G1) start from K-line KL1 and produce the sequence 11-12-13-14-15-16; (G2) start from K-line KL1 and produce the sequence 11-12-23-15-16; (G3) start from K-line KL2 and produce the sequence 21-12-23-15-16; (G4) start from K-line KL2 and produce the sequence 21-12-13-14-15-16. Note, solutions G1 and G3 are the same as S1 and S2, respectively, and therefore either fail or are not applicable. Solution G4 is also not applicable, since it starts from K-line KL2. However, solution G2 *is* applicable (since it starts from K-line KL1) *and* it will also *succeed*! Note, G2 is also the result of *combining* 2 K-lines, KL1 and KL2 and G2 is not available as an option if the two K-lines are not merged.

3.2 Application 2: Machine Learning

AKL can be useful in machine learning scenarios. We illustrate our case for a simple board game – Tic-Tac-To. The method can be applied to any strategy game, such as chess, go, etc.

The game of Tic-Tac-To (TTT) is played by 2 players, PX and PO, using a 3x3 board, as shown in Figure 8.

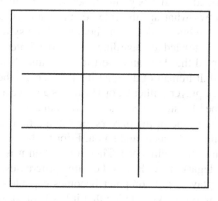

Fig. 8. Typical board for Tic-Tac-To game

Players PX and PO make alternating moves during which each player marks the board with a 'X' or an 'O'. Player PX uses always a 'X", while player PO uses always an 'O'. If any of the two players manages to mark the board with three 'X' or three 'O' to form a straight line (horizontal, or vertical, or diagonal), then that player wins. If the board has been completely marked and none of the two players achieves a 3-same-symbol straight line, then the game is a tie, i.e., none of the two players wins. Figure 9 shows two winning and one tie board configurations.

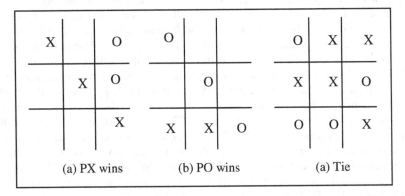

(a) PX wins   (b) PO wins   (a) Tie

Fig. 9. Two winning and one tie Tic-Tac-To board configurations

Next we describe how the AKL can be utilized to facilitate computer learning for the game of Tic-Tac-To, when one of the players is a computer, C, and the other player is a human, H. In our setting, C marks the board with the symbol 'C' and H marks the board with the symbol 'H'. If either C or H manage to achieve a straight line of 3-same symbols, (C, C, C or H, H, H) then the corresponding player wins; if the board has been completely marked and none of the two players achieved a straight line with its symbols, the game is a tie and a new game may begin. In our setting, we form a K-line for each game. The assumption is that many games are played, i.e., many K-lines are formed. At the beginning, before any game is played, we assume that the computer knows what are the rules of the game (i.e., it knows that a legal move is to mark any one empty slot of the board with its symbol 'C') but it does not possess any strategic knowledge regarding what constitutes a good move. On the other hand, it is assumed that H is an expert in this game. Note, for the simple game of Tic-Tac-To, any adult human of average intelligence can be considered an expert. The idea is that as each player (either C, or H) makes a move a K-node is formed. The K-nodes that are formed by successive moves are connected to form a K-line. The K-line ends with the win of one of the players, or when a tie occurs. Then a new game begins and a new K-line is formed, and so on. Prior to the first move of the very first game, the computer has no intelligence. Therefore, when it is its turn to move, it just makes a random (but legal) move. But as the game progresses, and especially as more and more games are played, the computer becomes capable of making more intelligent moves by drawing on the knowledge that it has been stored in the AKL that has been formed up to that point. The details of the K-node formation and the AKL formation are described next.

Event Structure. As mentioned earlier, each K-line consists of K-nodes and each K-node contains an event. The events of two consecutive K-nodes are related by causality, i.e., the event of the successor K-node is an immediate consequence of the event of the immediate predecessor K-node. In our setting, we define the event to be a move. Figure 10 shows two consecutive K-nodes K1 and K2 with their events $e^{\langle K1 \rangle}$ and $e^{\langle K2 \rangle}$.

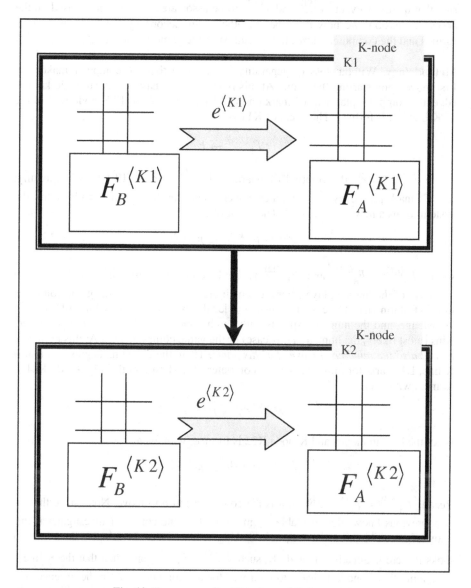

Fig. 10. Two consecutive K-nodes and their events

In Figure 10, each K-node K (K = K1 or K2), contains 2 Tic-Tac-To board configurations, $F_B^{\langle K \rangle}$ and $F_A^{\langle K \rangle}$. $F_B^{\langle K \rangle}$ denotes the Tic-Tac-To board configuration *before* the move during the formation of K-node K and $F_A^{\langle K \rangle}$ denotes the Tic-Tac-To board configuration *after* the move during the formation of K-node K. Note, for two K-nodes K1 and K2 to be connected such that K1 is the predecessor of K2, it must be that $F_A^{\langle K1 \rangle} = F_B^{\langle K2 \rangle}$. That is, the resulting TTT board in the predecessor K-node K1 must be the starting board configuration in the successor K-node K2. This is the criterion that qualifies events $e^{\langle K1 \rangle}$ and $e^{\langle K2 \rangle}$ to be associated by causality. Based on the above discussion, we now describe a methodology according to which an AKL is formed and the computer learns, in the context of the game of Tic-Tac-To.

Methodology. Without loss of generality, we assume that the computer makes the first move upon starting the game. At this point, without having any strategic knowledge, the computer places a 'C' mark on the TTT board and the 1st K-node K11 of the 1st K-line L1, is formed. The event of K11 is

$$e^{\langle K11 \rangle} : F_B^{\langle K11 \rangle} \to F_A^{\langle K11 \rangle}$$

where $F_B^{\langle K11 \rangle} = \varnothing$, the empty TTT board, and $F_A^{\langle K11 \rangle}$ is the TTT board containing the 'C' mark placed by the computer-player. Then, the human player plays and the system forms a new K-node, K12. The event of K12 is

$$e^{\langle K12 \rangle} : F_B^{\langle K12 \rangle} \to F_A^{\langle K12 \rangle}$$

where $F_B^{\langle K12 \rangle} = F_A^{\langle K11 \rangle}$ and $F_A^{\langle K12 \rangle}$ is the TTT board configuration resulting from the move of the human player. For the remaining of the 1st K-line, the game continues in this fashion (i.e., the computer makes a legal move without any kind of strategic knowledge, and the human responds with another move) until one of the two players wins (most likely, the human in this case) or the game comes to a tie. At this point, we *record also the number of moves for this game*. Then, the 2nd game begins and a new K-line, L2, starts forming. Again, the computer plays first and the 1st K-node K21 is formed, with event

$$e^{\langle K21 \rangle} : \varnothing \to F_A^{\langle K21 \rangle}.$$

Next, the human moves and K-node K22 is formed, with event

$$e^{\langle K22 \rangle} : F_B^{\langle K22 \rangle} \to F_A^{\langle K22 \rangle}.$$

Recall, $F_A^{\langle K21 \rangle} = F_B^{\langle K22 \rangle}$. Next, it is the computer's turn to move. Note, since there is now previous knowledge available (from K-line L1), the computer investigates if that knowledge can be utilized to help it in making its next move. In doing so the computer looks if there is already a K-node K, such $F_B^{\langle K \rangle} = F_B^{\langle K22 \rangle}$ and such that the K-line L containing that K-node leads to (i) a win for the computer, or (in the absence of a computer-winning K-line) (ii) to a tie, or (in the absence of (i) and (ii)) (iii) L is the longest K-line that leads to a human-win. In other words, the computer player tries to minimize its chances for defeat, and in doing so it looks for a move that – from past experience, lead to either a computer win, or (if that is not available) to a tie, or (if that is also not available) to a move that will postpone the defeat of the computer player the longest. Assume K_x is the K-node that is selected for the computer player to move, based on the above conditions/criteria. Then, the computer player moves by making the move dictated by the found K-node K_x. Note, at this point an *intersection* is formed in the AKL. Namely, the current K-line L2 intersects with the K-line containing the found K-node K_x. The game resumes, with the human player making the next move. After that, the computer player moves and its move is, again, based on the criteria described above – i.e., by finding (if available) a K-node that leads to an educated guess of postponing its defeat for the longest. The process continues, as described above, and new K-nodes are formed, as applicable, K-line intersections are also formed (as described above), and the AKL is taking shape, for as long as it is desired.

Testing. In testing our system, we judge the computer's learning progress by monitoring the amount of effort that is progressively required by the human to win. As the AKL becomes denser, it is expected that the computer player will choose K-nodes that belong to winning K-lines and, as a result, the moves that correspond to those K-nodes will make it harder for the human player to win. Of course, for the simple game of Tic-Tac-To, the human can always win, or, in the worst case, force a tie. However, for a more complicated game such as chess (or go), it is not at all clear that the human can keep winning after several games have been played. We expect that with enough training of an AKL, the computer can become a better expert in any of these complex games. We intent to test our method and report our findings, in a sequel paper.

Why is this way of machine learning interesting. Our proposed machine learning method (via the formation and utilization of an AKL) is novel and interesting because it does *not require that any knowledge is available up front*. The AKL can be formed progressively, as more games are played (i.e., more knowledge and experience is acquired), and there is also no limit of how much knowledge can be accumulated. As more knowledge and experience is acquired and captured into the AKL, the computer learner's performance is expected to improve. This, we argue, resembles the way that actual human beings learn. That is, as more experiences are acquired with the passage of time, humans, in general, perform better. Another interesting point in the nature of the learning facilitated by the AKL is that the *quality of learning is proportional to the quality of the "teacher"*. It is expected that an AKL whose K-node are formed by moves that are made by a not-so-intelligent (or, not-so-careful) human being, will be inferior to an AKL whose K-nodes are formed by moves that are made by a highly intelligent (or, skillful, or very careful) human being. As such, a machine that bases its moves on an inferior AKL is expected to underperform a machine that bases it moves on the superior AKL. Note, this analogy also resembles the real-life scenario in which, in general (all other things being equal), students of better teachers tent to become more skillful (i.e., learn more) than students of not-so-good teachers.

4 Comparison with Neural Networks

We provide a general comparison between the essence of AKL and ANN. Note, this is meant to be only a very general, although accurate, contrast between the proposed structure, AKL, and the long-existing well-known structure of ANN.

ANN is a structure that aids in decision making. After training, it answers, essentially, "yes"-"no" questions, upon presentation of input for a task. The task has to be within the domain that the ANN has been trained. That is, the ANN does not have general knowledge, neither does it have the ability for creativity and for combining multi-domain knowledge. In this sense, ANN is a *specialist (rather than a polymath) that continuously hones its skill to perform a certain task, without any creative abilities*. An ANN may increase its knowledge, but the newly acquired knowledge is *strictly confined* within the boundaries of performing better the *same* task that it used to perform before.

AKL is a structure that accommodates learning, but no decision making. An AKL does not point to a single piece of knowledge; on the contrary, it spans *several domains* of knowledge and this knowledge is expanded. The expansion is done with the incorporation of new K-lines into the existing AKL graph, regardless of the knowledge domain from where those K-lines come from. By incorporating more K-lines, AKL builds on existing knowledge and expands its ability of many different alternative "ways to think". As such, it facilitates creativity (i.e., allowing the generation of new "ways to think"), by allowing the formation of paths comprised from edges from *different* K-lines (as for example, cases (a) and (d) of Figure 4). The number of new "ways to think" is, in most cases, significantly greater than the number of the original "ways to think" that were used to form the AKL graph due to the periodic fun-outs that we encounter at K-line intersections. In this sense, AKL is the *creative polymath that continuously expands its knowledge and increases its chances for creative thinking*.

5 Conclusion

We present AKL, a novel approach in how to use K-lines. We illustrate via two applications of AKL that the proposed method can solve a problem that no other known method can solve as easily, as well as how an AKL can be used to facilitate a novel way of machine learning. It is hoped that the proposed structure is a potential tool for building intelligent systems, complementary to ANN.

Our future plans include refining the proposed method and investigating its potential for *machine learning* (such as the TTT example outlines here) and for *artificial creativity*. For *machine learning*, our immediate plan is to formulate a way to *automate* the testing of the TTT example described earlier. Note, testing of our methodology as described in Section 3 is possible, but it is cumbersome since it relies on continuous human participation. Since it is desirable that there is a *large* number of moves and games played in order for the AKL to acquire a *substantial* amount of knowledge that can be used by the machine, it is impractical to utilize an actual human for this purpose. Therefore, the input of a human participant has to be automated. Fortunately, for the TTT game outlines here, the *entire* knowledge space can be generated since this particular game is fairly small. We are currently working on formulating the problem of how this space can be utilized to automate the input of a human expert. In addition to the game of Tic-Tac-To, it is interesting to formulate the automation of human expert input for more complex games, such as chess and go. Another interesting research direction as an application of AKL, is the area of *artificial creativity*. Note, an inherent characteristic of AKL is to combine parts of K-lines and form new K-lines comprised of those parts. We argue that this is the essence of creativity, that is, the ability to form new "ideas" by using existing knowledge, or old ideas. In the context of the AKL, the newly formed K-lines represent the new ideas, whereas the existing knowledge is the sequence of segments of the existing K-lines that are used to form the new K-lines. In this sense, the AKL might be a suitable candidate structure for artificial creativity. It is our intention to delve into this line of research. One major obvious issue is how to judge the suitability of the newly generated K-lines, i.e., how to zero-in into *only* the meaningful generated ideas among all generated ideas, since the latter can be too many and, among those, many/most of them might not make sense. Note, again, the same phenomenon is encountered with natural creativity. That is, for every new idea that is generated by a person, there is always the issue of how sensible (or applicable) that idea is. We will be delighted if the AKL can shed some light to the artificial creativity problem, which is a many-decades old and unsolved problem in Artificial Intelligence.

References

1. Minsky, M.: K-Lines: A Theory of Memory. Cognitive Science 4, 117–133 (1980)
2. Minsky, M.: The Society of Mind. Simon & Schuster (1986)
3. Minsky, M.: The Emotion Machine. Simon & Schuster (2006)
4. Sloman, A.: Grand Challenge 5: The Architecture of Brain and Mind: Integrating Low-Level Neuronal Brain Processes with High-Level Cognitive Behaviours in a Functioning Robot. Technical Report COSY-TR-0607 (July 2006),
   http://www.cs.bham.ac.uk/research/projects/cosy/papers/
   #tr0607
5. Singh, P.: Failure Directed Reformulation, M. Eng. Thesis, Massachusetts Institute of Technology, Department of Electrical Engineering and Computer Science (March 10, 1998)
6. Singh, P.: EM-ONE: An Architecture for Reflective Commonsense Thinking, Ph.D. Thesis, Massachusetts Institute of Technology, Department of Electrical Engineering and Computer Science (June 2005)
7. Toptsis, A.A., Dubitski, A.: Iterative K-line Meshing via Non-Linear Least Squares Interpolation of Affectively Decorated Media Repositories. The Open Artificial Intelligence Journal 2, 46–61 (2008)

Spike Sorting Method Based on Two-Stage Radial Basis Function Networks

Jianhua Dai[1,2], Xiaochun Liu[1,2], Shaomin Zhang[1,3,4], Huaijian Zhang[1,3,4], Qingbo Wang[1,3,4], and Xiaoxiang Zheng [1,3,4]

[1] Qiushi Academy for Advanced Studies, Zhejiang University, Hangzhou 310027, China
[2] College of Computer Science and Technology, Zhejiang University, Hangzhou 310027, China
[3] College of Biomedical Engineering and Instrument Science, Zhejiang University, Hangzhou 310027, China
[4] Key Laboratory of Biomedical Engineering of Ministry of Education, Zhejiang University, Hangzhou 310027, China
jhdai@zju.edu.cn

Abstract. In this paper, 2-stage Radial Basis Function (RBF) Network method is used for neural spike sorting. Firstly, raw signals are obtained from Neural Signal Simulator, and added white noise ranged from -10dB to -40dB. Secondly, spikes are detected out with matched filter from signals. Lastly, 2-stage RBF networks are constructed and the spikes are sorted using RBF networks. The experiments show that 2-stage RBF network is an effective tool for neural spike sorting.

Keywords: RBF, spike sorting, neural network, K-means.

1 Introduction

Classification of neural action potentials is a prerequisite to study neural activity and brain functions. In extracellular recordings, a single electrode placed in the cortex usually records action potentials from several neurons nearby. To obtain the firing sequences of each neuron, spikes, which represent action potentials, has to be classified by some methods. The goal of spike sorting is to find the firing positions in time from each neuron. A lot of methods have been applied for spike sorting [1].

Radial basis function (RBF) networks are a type of artificial neural networks. It's proved that RBF networks have powerful function approximation capabilities, good local structure and efficient training algorithms. They have been successfully applied to many fields such as speech pattern classification, image processing, nonlinear system identification, nonlinear feature extraction etc. [2,3]. In this paper, we propose a method for neural spike sorting task based on 2-stage RBF networks. The experiments show that Radial Basis Function (RBF) Network is an effective tool for neural spike sorting task.

T.-h. Kim et al. (Eds.): FGCN 2008 Workshops and Symposia, CCIS 28, pp. 98–105, 2009.

2 Two-Stage Radius Basis Function Network

There are three layers in a typical RBF network, i.e. input layer, hidden layer and output layer. Input layer is made up of conception units, which receive input from outside. Hidden layer applies a nonlinear transformation between input layer and hidden layer. Output layer calculates the linear weighted sum of hidden units' output and provides the result after a linear trans-formation. The scheme of RBF networks is depicted in Figure 1.

It actually implements a mapping $f : R^n \rightarrow R$ as follows:

$$f(x) = \sum_{i=1}^{n} \omega_i \varphi(\| x - C_i \|) \tag{1}$$

Where $x \in R^n$ is the input vector, $\varphi(\square)$ is a transform function from Rn to R, $\| \square \|$ is

the Euclidean norm. $\omega_i, 1 \le i \le n$, are weights of each hidden unit.

$C_i \in R^n, 1 \le i \le n$, are called RBF centers, and n is the number of centers.

Because theoretical investigation and practical results suggest that the choice of $\varphi(\square)$ is not crucial to the performance of RBF network. In this paper, we adopt Gaussian function as $\varphi(\square)$.

$$\varphi(r) = \exp(-\frac{r^2}{2\sigma^2}) \tag{2}$$

Where σ is the width of the receptive field.

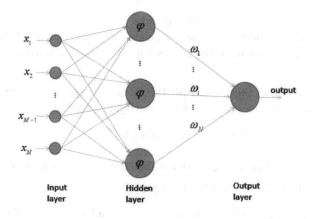

Fig. 1. Scheme of RBF network

So in training stage of RBF network, we should determine the RBF centers, $C_i \in R^n, 1 \le i \le n$, the width of receptive field, $\sigma_i \in R, 1 \le i \le n$, and the weights of each hidden unit, $\omega_i \in R, 1 \le i \le n$.

2.1 Training Stage of RBF Networks

In a typical two-stage RBF network, the mapping $f : R^n \to R$ is established in two stages. In the first stage, RBF centers are determined by K-means cluster, and then the width of the receptive field is calculated. In the second stage, the weights are determined by gradient descent algorithm.

Determine RBF centers using K-means method. K-means algorithm is a type of self-organizing algorithm. It performs the clustering task by iteratively decreasing distance between members of each cluster and the corresponding cluster centers. Here an improved K-means method is used to determine centers of RBF to capture more knowledge about the distribution of input patterns[4]. The method is illustrated as follows:

Firstly, we initialize cluster centers. We use 'Farthest points' as initial cluster centers. To get these farthest points, we have initially that $A = \varnothing$ and B is the whole data set to cluster. At first, two patterns which are farthest to each other are removed from B and added to A. Then iteratively move pattern x from B to A, whose minimum distance from any pattern in A is maximum. The iteration will continue until the number of patterns in A equals the intended number of clusters.

Secondly, we should determine which cluster each point belongs to. Calculate distance between each input vector and its corresponding cluster center. The input vector x is added to the cluster from whose center the distance is shortest. Then, update centers of each cluster using the following formula.

$$C_i = \sum_{x \in S_i} x \qquad (3)$$

Where $S_i, 1 \le i \le n$, is samples of the ith cluster.

At last, we should judge if termination conditions is satisfied. If all clusters remain the same after this iteration, or the maximum iteration number is reached, the iteration will terminate

Calculate the width of receptive field. The width of receptive field can be calculated according to distances between each RBF center gotten above.

$$\sigma_i = \lambda d_i = \lambda \min_j (\| C_i - C_j \|) \qquad (4)$$

Where λ is the overlapping coefficient. It is an important parameter of K-means RBF networks which effects how much the receptive fields overlap each other.

Determine the weights of hidden units using gradient descent algorithm. First, initialize the weights randomly. For each training sample, calculate its error from the corresponding target.

$$e_i = d_i - y_i = d_i - \sum_{j=1}^{N} \omega_j \varphi(\| C_i - C_j \|) \qquad (5)$$

Where d_i is the desired output of the ith training sample. y_i is the actual output of RBF network, which is linear weighted sum of hidden units' output.

$$e_i = d_i - y_i = d_i - \sum_{j=1}^{N} \omega_j \varphi(\| C_i - C_j \|) \qquad (6)$$

Then calculate the total error E.

$$E = \frac{1}{2} \sum_{i=1}^{P} e_i^2 \qquad (7)$$

where P is the number of the training samples. If E is less than the error limit, the procedure will terminate, else update the weights, and calculate the error again.

$$\omega_i \leftarrow \Delta \omega_i + \omega_i \qquad (8)$$

$$\Delta \omega_i = -\eta \frac{\partial E}{\partial \omega_i} = \eta \sum_{j=1}^{P} e_j \varphi(\| X_j - C_i \|) \qquad (9)$$

Where η is learning rate, which affects the convergence speed of the algorithm, X_j is the jth training sample.

2.2 Testing Stage of RBF Networks

When using the ith sample as input vector, output of the jth hidden unit is

$$\varphi_{ij} = \varphi(\| X_i - C_j \|), \, i = 1, 2, ..., P; \, j = 1, 2, ..., N. \qquad (10)$$

So the output matrix of hidden layer is

$$\Phi = \left[\varphi_{ij} \right]_{P \times N} \qquad (11)$$

The weights matrix between hidden layer and output layer is

$$W = [\omega_1, \omega_2, ..., \omega_N] \qquad (12)$$

Then the output vector of RBF network is

$$F(X) = \Phi W \qquad (13)$$

Actually, after training we get the weights matrix W and the function $\varphi(\square)$. Then using X as the input of RBF networks, we can get the results by the above formula.

3 Data Source

3.1 Obtain Raw Signals and Add White Noise

To evaluate the network in a range of noise intensity, experiments have been down with synthetic signals. The signals used in the paper are from the Neural Signal Simulator. The circuit simulates the output of a Cyberkinetics 100-electrode array. Each virtual electrode is simulated as detecting microvolt signals from three separate neurons located at different distances from the recording site. The amplitudes of the three action potentials as well as the kinetics of their responses differ in a manner consistent with real world signals. In the signal simulator, Local Field Potentials are also run. The simulation on each output channel consists of a sequence of three individual action potentials that 'fire' one after the other at a 1s intervals, figure 1. This firing sequence repeats nine times. Then, every 10 seconds, a one second burst of activity is simulated. The burst consists of the same train of three individual action potentials, and they are repeated with an inter action potential interval of 10 milliseconds. The output impedance of each of the 128 output channels is 220k. This simulates the output impedances of the electrodes of the Cyberkinetics array. The amplitudes of the three action potentials are 300, 250, and 200µV.

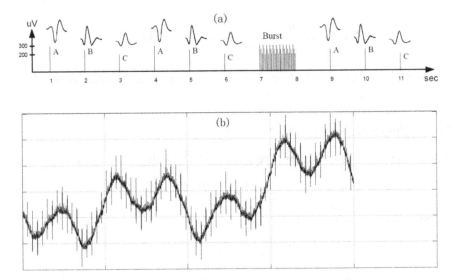

Fig. 2. Output of Neural Signal Simulator. (a) Specifications of the Action Potential from the Simulator. In every 10 seconds, a one second burst of activity is simulated, which consists of the same train of three individual action potentials, and they are repeated with an inter action potential interval of 10 milliseconds. The amplitudes of the three action potentials are 300, 250, and 200µV. (b) Waveforms of the Raw Signal from the Simulator. Pinnacles in the waveforms represents spikes that recorded by the electrode. The waveform contains local field potential as well, in low-frequency, one kind of disturb in the action potential analysis, which can be eliminated by differential treatment.

Raw signals are superposed with white Gaussian noises in the measurement of dB, which is the scalar of SNR that specifies the signal to noise ratio for better simulation of real situation, using matlab function awgn(x, SNR). The scalar SNR specifies the signal-to-noise ratio per sample in dB. This syntax assumes that the power of x is 0 dBW, and "dBW" representing a power level in decibels relative to 1 Watt.. Then we can achieve the spike classification results in a range of different degrees of the environmental noise.

3.2 Spike Detection

The typically used technique for detecting the action potentials is amplitude threshold crossing, which extracts the action potentials when the amplitude exceeded a threshold derived from the standard deviation of the signal from the baseline. Methods such as matched filter are also widely used to work as a detector for action potentials, which might achieve better performance than the amplitude threshold crossing in the noisy environment [5]. Here, we use the method of matched filter as a detector that maximized the SNR of the signal by a system whose transfer function is the inversion of the typical action potential waveforms, which works better in our experiments.

3.3 Spike Sorting

After spike detection, we get waveforms which are 90 points in length. According to the waveforms obtained in experiment, we choose a part of 19 points as input vector of RBF network (Fig. 3). To make train signals distribute uniformly in the data set, first we use K-means to calculate centers of each class of spikes; then sort the vectors according to its distance to the corresponding class center. From the data set, we get one training vector, then one testing vector, then one training vector, etc. In this way we get about 750 groups of training samples and 750 groups of testing samples for each SNR. After training, we sort the spikes using the trained RBF network.

4 Experiment Results

The correct ratio of spike sorting under different SNR are shown in Table 1. Tab. 1 shows that when SNR is higher than -25 the correct ratio is higher than 95%. After that, the correct ratio decrease fast and reaches about 70% when SNR = -40dB.

Table 2 shows a comparison between RBF network and BP network. From the table we find the correct ratio of RBF network decrease earlier (when SNR = -30) than BPNN (when SNR = -35). But RBF network achieve a higher correct ratio when SNR reaches -35dB or even lower.

Table 1. Correct ratio of RBF networks when SNR of test signals decreasing. To converge in reasonable time at a relatively low SNR, we have to larger the tolerance of the training period. Because the neural network has good generalization ability, it's not necessary to train the network for every group of data. So the train sets are only made up with samples of -10dB, -30dB, -35dB.

Test Signals(dB)	Correct Ratio(%)	Tolerance	Train Signals(dB)
-10	98.67	0.5	-10
-15	99.33	0.5	-10
-20	99.0	0.5	-10
-25	96.33	8	-30
-30	86.67	8	-30
-35	75.67	60	-35
-40	73.67	60	-35

Table 2. Comparison of the correct ratio using RBF networks and BP networks. The results show that RBF networks achieve a better result when SNR comes to -35dB or lower.

Test Signals(dB)	K-means RBFNN(%)	BPNN(%)
-10	98.7	99.3
-15	99.3	98.7
-20	99.0	98.7
-25	96.3	96.0
-30	86.7	90.0
-35	75.7	69.3
-40	73.7	71.3

5   Conclusion

In this paper we propose 2-stage RBF network method for neural spike sorting. The results show that RBF network performs well under high noise. So RBF network can be an effective tool in spike sorting. However, there are some problems with this method: Some parameters such as width of reception field need to be set manually, which is a time-consuming task. The performance is affected greatly by distribution of samples because of the application of K-means method in the first stage of RBF network. Therefore, in our future work we will consider improve the performance using Genetic Algorithm to make RBF network self-adaptive [6]. We will do further research on neural decoding in Brain-Computer Interface [7].

Acknowledgements

This work is supported by the National Science Foundation of China (60703038, 30800287), the Excellent Young Teachers Program of Zhejiang University and the Research Foundation of Center for the Study of Language and Cognition of Zhejiang University.

References

1. Lewicki, M.S.: A review of methods for spike sorting: the detection and classification of neural action potentials. Network: Comput. Neural Syst. 9(4), R53–R78 (1998)
2. Mulgrew, B.: Applying radial basis functions. IEEE Signal Processing Magazine, 50–65 (1996)
3. Chen, S., Cowan, C.F.N., Grant, P.M.: Orthogonal Least Squares Learning Algorithm for Radial Basis Function Networks. IEEE Transactions on Neural Networks 2(2), 302–309 (1991)
4. Sing, J.K., Basu, D.K., Nasipuri, M., Kundu, M.: Improved K-means Algorithm in the Design of RBF Neural Networks. In: TENCON 2003. Conference on Convergent Technologies for Asia-Pacific Region, vol. 2, pp. 841–845 (2003)
5. Pfurtscheller, G., Fischer, G.: A new approach to spike detection using a combination of inverse and matched filter techniques. Electroencephalogr. Clin. Neurophysiol. 44(2), 243–247 (1978)
6. Chen, S., Wu, Y., Luk, B.L.: Combined genetic algorithm optimization and regularized orthogonal least squares learning for radial basis function networks. IEEE Transactions on Neural Networks 10(5), 1239–1243 (1999)
7. Dai, J.H., Liu, X., Zhang, S., Zhang, H., Yi, Y., Wang, Q., Yu, S.Y., Chen, W., Zheng, X.: Analysis of neuronal ensembles encoding model in invasive Brain-Computer Interface study using Radial-Basis-Function Networks. In: Proceeding of the 2008 IEEE International Conference on Granular Computing, pp. 172–177. IEEE Press, Los Alamitos (2008)

The Establishment of *in vitro* Model of Primary Rat Brain Microvascular Endothelial Cells

Jun Li[1], Liang Peng[1], Xiao-Lu Liu[1], Hong Cao[1,*],
Sheng-He Huang[1,2,*], and Chun-Hua Wu[2]

[1] School of Public Health and Tropical Medicine, Southern Medical University,
Guangzhou, Guangdong 510515, China
[2] Children's Hospital Los Angeles, the University of Southern California,
Los Angeles, USA
{Jun Li,Liang Peng,Xiao-Lu Liu,Hong Cao}gzhcao@fimmu.com,
{Sheng-He Huang,Chun-Hua Wu}shhuang18@hotmail.com

Abstract. In order to establish an *in vitro* brain microvascular endothelial cells (BMEC) model of blood-brain barrier (BBB), we isolated and cultivated primary microvascular endothelial cells from grey matters of rat brain. The grey matter was squashed into homogenate, centrifugalized with dextran, and digested. Then the microvessels were seeded onto collagen-coated 25cm^2 tissue culture flasks. Then the cell was examined with microscope. Microscopically the isolated cells in the fusiform, when cultured in 24 hour, formed the cobblestone shape. These cells reached confluence in the form of uniform monolayer after 7 days in culture. Therefore, the primary BMECs of rat brain were successfully isolated and cultivated by using the above method, and we could use this model to study the relationship between BMEC and pathogenic microbes.

Keywords: Rat, Primary culture, BMEC.

1 Introduction

The first successful isolation of brain microvessels from animal brain was 30 years ago, and then many scientists studied several new ways to develop *in vitro* models system of the blood–brain barrier (BBB), in order to establish a more stable and similar to the blood–brain barrier *in vivo*. And from then on, these BBB model systems *in vitro* have contributed considerably to the understanding of the physiology, pharmacology, pathophysiology and centre nervous system infection of the BBB.

This Brain microvascular endothelial cell (BMEC) is the main component of blood-brain barrier (BBB). The BMEC monolayer *in vitro* model system are morphologically similar to endothelial cells of the BBB *in vivo*, characterized by tight intercellular junction, few pinocytic vesicles, and lacking fenestra. Markers for cells of endothelial origin, for example, angiotensin-converting enzyme, and factor VIII antigen, have been also demonstrated in this model[1]. The transfer function studies on the model *in vitro* showed that

---

* Corresponding authors.

T.-h. Kim et al. (Eds.): FGCN 2008 Workshops and Symposia, CCIS 28, pp. 106–114, 2009.
© Springer-Verlag Berlin Heidelberg 2009

whether passive or active transport dispersion process is found similar *in vivo*[2]. Because of these characteristics, this model system possesses many advantages over conventional *in vivo* studies, for example, rapid assessment of invasion of pathogenic microorganism and proteins, genes, cell signaling pathways which have business with bacterium invasion. This *in vitro* BBB model system would have a wide application for microbial meningitis and other studies about how pathogenic microbe destroy or pass through the BBB without expensive animals studies.

There were many discussed methods for isolations of BMEC were reported in the past three decades, and the most common means of isolation include homogenate, filtration, centrifuge and digestion. Besides, there were labs use tissue culture and white matter of brain for culture[3,6,9~15]. However, spending long time, big workload demanding instruments, survival difficulties of BMEC and growth of mixed fibroblasts are still common problems. How to cultivate endothelial cells with high purity and characteristics of the BBB, is always a difficulty for researchers no matter interior or abroad.

A method about separation and cultivate successfully for the brain microvascular endothelial cells with high purity and growth activity would be reported by this article.

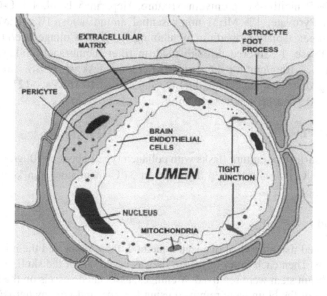

Fig. 1. The BBB structure: Brain microvascular endothelial cell is the main component of the BBB. The BBB *in vivo* is characterized by tight intercellular junction, few pinocytic vesicles, and lacking fenestra. (Source: Sheng-He Huang and Ambrose Y. Jong. Cellular mechanisms of microbial proteins contributing to invasion of the blood–brain barrier).

2 Materials and Methods

2.1 Animals

Eight rats about two- to three-week-old.

2.2 Reagents

RPMI-1640, DMEM containing 4000mg/L glucose and FBS were obtained from Gbico; L-Glutamine (200mmol/L), Penicillin-Streptomycin Mixture (10KU/L), 25% Trypsin – EDTA mixture were obtained from Genom; Dextran (clinical grade, average mol. wt. 70,000) were obtained from Seebio; DNase I , Heparin(10KU/ml PBS) were obtained from Sigma; Collagen Type I (100mg/28.65ml) were obtained from Upstate; Sodium Pyruvate (100mmol/L), MEM non-essential amino acids, MEM vitamins were obtained from Irvine; Nu-Serum were obtained from B&D; Endothelial cell growth factor (ECGF), and Collagenase/ Dispase were obtained from Roche.

2.3 Culture Solution

Isolation medium: DMEM containing 2%FBS and 5% Penicillin-Streptomycin Mixture; Growth medium: RMPI 1640 containing 10% FBS, 10% NuSerum, ECGS 30mg/L, 1% Penicillin-Streptomycin Mixture, Heparin 5 U/ml, 1% L-Glutamine, 1% Sodium Pyruvate, 1% MEM non-essential amino acids, 1% MEM vitamins; Digestive juice: Isolation medium containing 1mg/ml Collagenase/Dispase and 0.1mg/ml DNase I , filtered through 0.22μm filter membrane; Collagen Type 1: Diluted it to 5 ~ 10μg/cm^2 for the surface of culture flask with PBS. 30% Dextran: Dextran 60g dissolved in 200ml ddH$_2$O, autoclaved (0.15KPa, 121℃, 15min), and then added 20ml PBS in dextran solution.

2.4 Culture Flask

Covered the surface of culture flasks with collagen type I about 5~10μg/cm^2, and kept the flasks at 4℃ overnight. Then dry them in 37℃ incubator, then kept at 4℃.2.5 Isolation method.

2.5 Method

Rats were killed using cervical dislocated method, then, immersed them in 75% ethanol for 5 min. Then each brain of these rats was eased out of the skull, and the cerebellum and brain stem were removed. Meninges, associated vessels on the surface and white matters of the brain were removed using forceps, and gray matter (Figure 2.) of the brain was rolled slowly and carefully on a sterile filter paper. The isolated cerebral cortex was placed into 25ml isolation medium.

Then, gray matters were passed through a common metal net gently to pulverize them into smaller pieces. Translated the organization levitation medium into homogenizer, homogenized the medium till it looks like milk shake. The homogenized samples were transferred into a 50ml centrifuge tube. Then 25ml 30% dextran solution was added into this centrifuge tube. The mixture was centrifuged at 4500g, 4℃ for 20 min. Discarded supernatant and resuspended the pellets with PBS, and centrifuged at 150g, 4℃ for 5 min to wash dextran away. Then the step repeated twice.

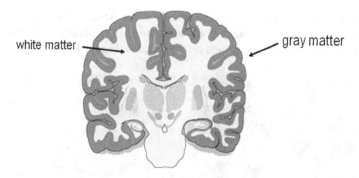

Fig. 2. Gray matter was used to isolate the brain microvascular endothelial cells in the experiment

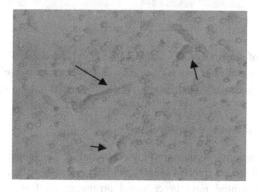

Fig. 3. The isolated microvessels after digesting: digested using Collagenase/dispase and DNase I at 37℃ for 30min, till most microvessels beaded and shorter then before. (*200 ✕*)

The pellets was resuspended and dissociated with 2ml digestive juice, at 37℃ for 30min, or the digestive suspension would be observed using microscope every 10min from 20th minute on, till most microvessels beaded and shorter then before (Figure 3). The microvessels were centrifuged at 150g, room temperature for 5min, and discarded supernatant. Then the pellet was resuspended with PBS and centrifuged at 150g, room temperature for 5min, twice. Then the pellet was resuspended with culture medium, plated in collagen-coated 25cm^2 flask (Greiner, Bio-one, Germany), 1.5ml/flask, incubated in a water-saturated atmosphere of air and 5%CO_2. 24 hours later, when the cells had adhered, the culture medium should be changed every other day, until the cells became a confluent monolayer. Then the cells could be freezing or passage.

3   Results

3.1 BMEC Were Cultivated in 24 Hours

The collagen-coated surface is very beneficial for adherence of microvessels, micro-scopically capillaries and a few red cells could be observed. After 5 to 6 hours of culture, some microvessels had adhered (Figure 4, A). About 12 hours later, the most microvessels had attached. Twenty hours after the vessels planted, cells had climbed out and proliferated, gradually formed cell aggregates (Figure 4, B).

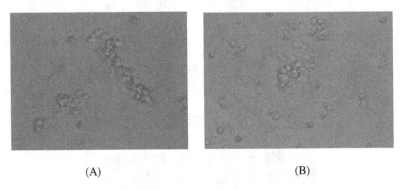

(A)   (B)

Fig. 4. The growth of primary BMECs within 24 hours cultivated. (A) The digested microvessels had adhered after cultivated five hours; (*200 ✕*); (B) The microvessels in flask after incubated 20 hours, cells had climbed out, and then gradually formed cell aggregate. (*200 ✕*)

3.2 BMEC after 1 Week of Culture

Cultivated in the medium containing Nu-serum and ECGF longer than 24 hours, the cells climbed out from the microvessels increased obviously. The cells are fusiform or polygonal, and presented monolayer spiral arrangement, formed a cell aggregate. These cell aggregates proliferated gradually, would reach confluence at the 7th day, and presented "road-metal" arrangement (Figure 5).

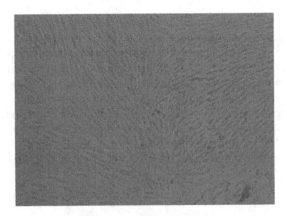

Fig. 5. The BMEC after 7 days' culture, the cell aggregate proliferate gradually, the cells reached confluence after 7 days, and presented "road-metal" arrangement. (*40 ✕*)

4 Discussions

When Panula and other people first proved that BMEC can be cultivated by tissue culture method, including that primary and genetic individual BMEC had been applied in the following future experimental external BBB model system. Compared with the individual microvessel which is the earliest *in vitro* model system of the BBB, its main advantage is that it can be applied to research the transport process of the transversal cells under the proper experimental conditions. At the mean time, the external experiments of central nervous system infection related to invading BMEC is also mainly dominated by such model.

A method of isolation and culture BMECs effectively but relatively low requirements for the instrument was explored. BMEC could be separated without high-speed centrifuge, and this method consumes less time, too.

Comparing other separation methods interiorly and abroad, the step of passing the gray matter through the metal gauze instead of the fine pruning, and tissue suspension obtained after this grinding process was more uniform, the pieces were smaller, the operation is more finely and timesaving. The use of tissue homogenizer which operated gently and slowly can crush tissue with less damage of cells, and the suspension can be stratified by centrifugation directly. Compare with twice digestions method, this process result in shorter operation time, and less complicated steps. But the mechanical damage to microvessels would be more severe than digestion's if operated roughly, which even affect the activity of cells. So this process must operate gently and in low temperature. In addition, the use of screen mesh suggested by many labs was attempted and tested. It is considered that screen mesh can reduce the immixing with blood cell effectively, but the function of filtering off gliacyte and astrocyte is not obvious, oppositely, the loss of blood vessels in the process increased. If meningeal fragments were remained apparently because of rough operation when stripping meninge, we can filter it once using the screen mesh with 150 micron aperture after homogenate to remove meningeal fragments.

Dextran zonal centrifugation can get good isolation effect. According to previous reports, we attempted the step of centrifuge at different centrifugal speed and time [1, 3~5], including 7000g, 10min; 3500g, 20min; and 4500g, 10min, repeated three times. Of the three plans, the first one have higher centrifugal force which can isolated cells effectively, and also have short contact time between microvascular fragments and dextran, but it demand high requirement to equipment, and the high centrifugal force would decline the growth activity of vascular fragments. According to the culture experience, many microvascular fragments can't adhere to culture surface and grow slowly which usually show growing ability after have been cultivated between 24 to 48 hours. The second plan asks lower centrifugal force and keeps high microvascular growth activity, but microvascular can still found in upper layer of liquid and there are too many impurities in vascular layer because of the incomplete isolation. This plan can be taken if there were equipments for percoll operation for further isolation. The third plan use the method of repeated centrifugation to remove impurities as possible, but repeated centrifugation and long contact with dextran are the inhibitory factors of growth activity of cells, those steps also increased the loss of vascular fragment. Thus, according to these protocols we take a compromise centrifugal rate and centrifugal time which can isolate microvascular fragments well and show well pick-up rate of microvessels by once centrifugation and have less inhibition to cell growth activity. Besides, many labs use BSA which is considered better than dextran in this process. But according to our attempt and test, the result of culture using BSA is not much better than that of dextran. As to the ingredient of culture fluid, in the absence of Nu-serum or ECGF circumstances, we can use only one of them for our culture. But using both of them can promote the cell growing better. The planting density should be increased if only one of them to be used, the proper density is three culture flasks for BMEC from eight rats.

The growing ability was restrained obviously by miscellaneous cells, especially fibroblast after passage, thus BMEC need to be purified after passage. Several methods of purification such as digestion lag separation, adherence lag separation, miscellaneous cells curettage method, monoclonal culture and suppression medium culture methods were tried to remove fibroblasts. Digestion lag separation method and monoclonal culture method do not apply to remove fibroblasts. Curettage method could kill fibroblasts effectively, easily, but only partly. Because of adherence between BMECs and fibroblast, curettage fibroblasts often embroiled the endothelial cells around. Because medium D-Viline MEM can lead to fibroblast's growth inhibiting, which inhibit endothelial cells' growth as well, so cultivating the primary cell with D-Viline MEM when dominant growth of endothelial cells appeared can inhibit the growth of fibroblasts effectively and won't restrain the growth activity of endothelial cells which own the overall advantages.

The cells we cultured were fusiform or polygonal, and presented monolayer spiral, formed a cell aggregate. These cell aggregate presented "road-metal" arrangement when they reached confluence (Figure5). These characters are the same as reported before.

It has begun to attempt to cultivate endothelial cells abroad since 1960s, it mainly drew materials from the cerebral cortex tissues of mice, rats, cattle, pigs, goats, dogs and human. The source of rat brain is plentiful, it is relatively easy to perform sterile operation and draw materials, but the gray matter is comparatively little, it needs a large number of animals at one time. The gray matter of cattle brain is plentiful, it is easy to obtain comparatively pure gray matter tissue, but the source of brain tissue is relatively difficult, the cost is comparatively high, it is comparatively difficult to draw materials and it is easily to be polluted, the domestic and abroad documents of isolation and culture methods are quite different, the cultivated cell shapes are also different, so it seldom uses cattle as sampling animals in China. After the isolation technique of brain microvascular endothelial cells of animals becomes more and more mature, human's will also be isolated and cultivated, all the founded indexes of *in vitro* blood-brain barrier is closer to the blood-brain barrier inside human's body. But compared to the experimental animals, the source of human's healthy brain tissue is less, it is more difficult to draw materials and perform cells isolation and culture. In contrast, the source of rats is plentiful, which contain multiple kinds of mature pathological models, so rats are always usual sampling objects.

In fact, pigs and cows have been used to isolate brain microvascular endothelial cells in our laboratory, too. In our opinion, these experiments appeared well result of culture, but these animals are costly, and the operation of taking out brain is more difficult, and the brains of animals are much easier to be polluted by microbe. And we find that infancy animals' brain microvascular endothelial cells showed more excellent ability of proliferation than adult ones, it is more likely to succeed if we choose infancy animals. So, rats younger than 2 weeks could be used, but the assist of microscope to remove meninges is necessary. And laboratory has attempted to use the experimental miniature pigs after born about one month as sampling animals, the brain tissue volume is big, cerebral cortex is plentiful, and the animal body type is comparatively small, it is relatively easy to draw materials, the cost is cheaper than cattle brain, so the effect is comparatively satisfying, it is also a comparatively ideal experimental object.

Through the development in the past 30 years, there were several model systems that were constructed, which included the model of the blood-brain barrier (BBB) *in vitro* that is mainly constitute by the brain microvascular endothelial cell. The methods of the model construction and composition are different from each other, so the differences also existed in the results that given by different physiological and biochemical characteristics experiments, the model systems also have their own advantages and disadvantages. At present, there is no such particular model is better than other ones. Therefore, the research on blood-brain barrier usually will use two or more different models combined together.

Primary monolayer BMEC is an *in vitro* model system of the BBB for studying the physiology, pathology, pharmacology, central nervous system of infection, and also the first step of building the more advanced model of the BBB *in vitro*. With the means of isolation continuous improvement, BMEC and this *in vitro* model system of the BBB will get more extensive and be more in-depth studied.

Acknowledgments. This work was supported by Guangzhou Key Technology R&D Program (No.2008Z1-E401 to H. Cao) and the NIH (RO1 AI40635 to S.H. Huang).

References

1. Audus, K.L., Borchardt, R.T., et al.: Bovine brain microvessel endothelial cell monolayers as a model system for the blood-brain barrier. J. Annals of the New York Academy of Sciences 507(Biological), 9–18 (1987)
2. Akiyama, H., Kondoh, T., Kokunai, T., et al.: Blood-brain barrier formation of grafted human umbilical vein endothelial cells in athymic mouse brain. J. Brain Res. 858, 172–176 (2000)
3. Franke, H., Galla, H.-J., Beuckmann, C.T., et al.: Primary cultures of brain microvessel endothelial cells: a valid and flexible model to study drug transport through the blood–brain barrier in vitro. J. Brain Research Protocols 5, 248–256 (2000)
4. Xu, X.-F., Li, R.-P., Li, Q., et al.: Isolation and Primary Culture of Rat Cerebral Microvascular Endothelial Cells. J. Chinese Journal of Cell Biology 27(1), 84–88 (2005)
5. Ichikawa, N., Naora, K., Hirano, H., et al.: Isolation and primary culture of rat cerebral microvascular endothelial cells for studying drug transport in vitro. J. Iwamoto Journal of Pharmacological and Toxicological Methods 36, 45–52 (1996)
6. Yuzhen, Z., Wen, W., Yeping, T., et al.: Primary culture and Purification of mouse cerebral microvascular endothelial cells. J. Chinese Journal of Anatomy 30(5), 530–533 (2007)
7. Perrière, N., Demeuse, P.H., Garcia, E., et al.: Puromycin-based purification of rat brain capillary endothelial cell cultures. Effect on the expression of blood–brain barrier-specific properties. J. Journal of Neurochemistry 93, 279–289 (2005)

8. Jie-xiao, L., Chuan-qiang, P., Yan-lei, H.: In vitro isolation and culture of primary blood-brain barrier endothelial cells by double filtering technique. J. Journal of Clinical Rehabilitative Tissue Engineering Research 12(21), 4123–4126 (2008)

9. Jong, A.Y., Wu, C.-H., Jiang, S., et al.: HIV-1 gp41 ectodomain enhances Cryptococcus neoformans binding to HBMEC. J. Biochemical and Biophysical Research Communications 356, 899–905 (2007)

10. Xianghui, L., Luxiang, C., Xianghui, L., et al.: Culture of rat brain microvascular endothelial cells. J. Acta Academiae Medicinae Militaris Tertiae 29(20), 2011–2013 (2007)

11. Zhenqiang, S., et al.: Cell Culture (Revised editior). The world books publishing house, Xi'an (2004)

12. McCarthy, K.D., Vellis, J.D.: Preparation of separate astroglial and oligdendroglial cell cultures from rat cerebral tissue. J. Cell Biol. 85, 890–902 (1980)

13. Guo, Y., Jiang, L., Liu, G.-X., et al.: Expression of Drug Transporters in Primary Rat Brain Microvascular Endothelial Cells. J. Chinese Journal of Cell Biology 29, 560–564 (2007)

14. Mischeck, U., Meyer, J., Galla, H.J.: Characterization of γ-GT activity of cultured endothelial cells from porcine brain capillaries. Cell Tissue Res., 221–256 (1989)

15. Phillips, P., Kumar, P., Kumar, S., et al.: Isolation and characterization of endothelial cells from rat and cow brain white matter. J. Anat. 129, 261–272 (1979)

16. Ming-guang, Z., Tao, T., Yong-zhong, G., et al.: Culture of human cerebral capillary endothelial cell by separation of capillary fragment and the observation of vascular endothelial growth factor gene expression and cell ultrastructure. J. Chinese Journal of Clinical Rehabilitation 9(21), 211–213 (2005)

17. Calabria, A.R., Weidenfeller, C., Jones, A.R., et al.: Puromycin-purified rat brain microvascular endothelial cell cultures exhibit improved barrier properties in response to glucocorticoid induction. J. Journal of Neurochemistry 97, 922–933 (2006)

A FC-GSEA Approach to Identify Significant Gene-Sets Using Microarray Gene Expression Data

Jaeyoung Kim[1] and Miyoung Shin[2,*]

[1] Graduate School of Electrical Engineering and Computer Science, Kyungpook National University, Daegu, South Korea 702-701
widebrowboy@gmail.com
[2] School of Electrical Engineering and Computer Science, Kyungpook National University, Daegu, South Korea 702-701
shinmy@knu.ac.kr

Abstract. Gene set enrichment analysis (GSEA) is a computational method to identify statistically significant gene-sets showing differential expression between two groups. In particular, unlike other previous approaches, it enables us to uncover the biological meanings of the identified gene-sets in an elegant way by providing a unified analytical framework that employs a priori known biological knowledge along with gene expression profiles during the analysis procedure. For original GSEA, all the genes in a given dataset are ranked by the signal-to-noise ratio of their microarray expression profiles between two groups and then further analyses are proceeded. Despite of its impressive results in previous studies, however, the gene ranking by the signal-to-noise ratio makes it hard to consider both highly up-regulated genes and highly down-regulated genes at a time as significant genes, which may not reflect such situations as incurred in metabolic and signaling pathways. To deal with this problem, in this article, we investigate the FC-GSEA method where the Fisher's criterion is employed for gene ranking instead of the signal-to-noise ratio, and evaluate its effects made in Leukemia related pathway analyses.

Keywords: significant pathway, gene set enrichment analysis, gene ranking, Fisher's criterion, microarray data analysis.

1 Introduction

Recent advance in microarray gene expression studies has played an important role in elucidating and understanding complicated biological phenomenon occurred in various organisms by performing computational analysis on gene expression profiles along with a variety of biological resources [1], [8], [9]. In particular, one of the important challenges in gene expression studies is to identify differentially regulated genes between two groups and understand their biological meanings. For this purpose, the gene set enrichment analysis (GSEA) [1] has been lately developed as a unified analytical framework that employs a priori known biological resources with gene

---

* Corresponding author.

T.-h. Kim et al. (Eds.): FGCN 2008 Workshops and Symposia, CCIS 28, pp. 115–128, 2009.
© Springer-Verlag Berlin Heidelberg 2009

expression profiles. Unlike other previous approaches where gene expression profiles only are employed for identifying differentially regulated genes and then some additional steps are required for understanding their biological meanings, the GSEA approach combines the analysis step for finding differentially regulated genes and the interpretation step for understanding their biological significance into an integrated framework in a systematic manner. The usefulness of this approach in a variety of contexts has been shown in several previous studies [2, 3]. In spite of their impressive results, however, the gene ranking by the signal-to-noise ratio (SNR)[1] used for original GSEA produces some limitations such that for a specific gene-set it selects as significant ones either highly up-regulated genes only or highly down-regulated genes only in the end, not both at the same time. Thus, this makes it hard to reflect the situations in which both highly up-regulated genes and highly down-regulated genes play an important role, just like biological pathways.

To deal with this problem, in this paper, we investigate the method, FC-GSEA, to employ Fisher's criterion (FC) [5, 6, 7] for gene ranking in the gene set enrichment analysis. With the FC-based gene ranking, we make an attempt to discover significant pathways which may not be found by original GSEA using the SNR-based gene ranking.

This paper is organized as follows. Section 2 reviews and summarizes original GSEA approach and its uses in various applications. Section 3 describes our proposed approach, FC-GSEA, which employs the FC-based gene ranking method. Section 4 describes the experiments on Golub's leukemia dataset and also evaluates the results in a comparative manner with original GSEA. Finally, Section 5 concludes the paper with some discussions.

2 Gene Set Enrichment Analysis

As a computational technique to identify statistically significant gene-sets showing differential expression between two groups, the GSEA approach has attained a lot of attentions by many researchers, showing impressive results in a variety of applications [1, 2, 3]. Specifically, Subramanian et al. [1] demonstrated how the GSEA could yield some deep insights into several cancer studies such as leukemia and lung cancers. Also, E. Taskesen [2] used the GSEA for performing some cancer studies on human mammary epithelial cell lines, and in [3], the authors employed the GSEA and other computational methods to identify robust subtypes of diffuse large B-cell lymphoma and understand more effective treatment strategies.

A summary of original GSEA approach [1, 2] can be given as follows:

[STEP 1] Compute the enrichment score (ES) of gene-set

- Rearrange all the genes in the decreasing order of the signal-to-noise ratio between two groups
- Build candidate gene-sets by using a variety of biological resources
- For each gene-set, calculate Kolmogorov-Smirnov score on the ordered entire gene list and take the absolute maximum ES.

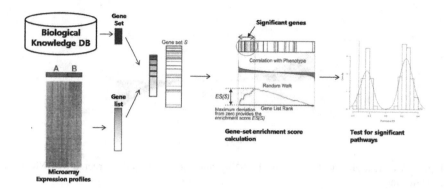

Fig. 1. A summary of gene set enrichment analysis approach [1]

[STEP 2] Estimate statistical significance of the computed ES

- Generate k different expression datasets by performing k random permutations on class labels of gene expression profiles.
- Perform the computation of ES values on the generated k different datasets and obtain a null distribution of ES.
- Use the null distribution to calculate a nominal p-value of the ES obtained in STEP 1.

[STEP 3] Adjust significance level for multiple hypothesis tests

- For each gene-set, obtain a normalized-ES by normalizing the computed ES to account for the gene-set size.
- Adjust false positive rate by calculating false discovery rate (FDR) for each normalized-ES and finally determine statistically significant gene-sets

3 FC-GSEA : Fisher's Criterion Based GSEA

The FC-GSEA method is a gene-set enrichment analysis approach to identify significant gene-sets showing statistically significant differences in gene expression intensities between two classes, where the Fisher's criterion is employed for gene-ranking. The detailed description of FC-GSEA is given as below.

3.1 Motivation

In the original GSEA approach [1, 2], the statistical significance of gene-sets is estimated based on the enrichment score (ES) calculated for each gene-set. Also the ES computation is made by calculating the Kolmogorov-Smirnov(KS)score with the ordered entire gene-list generated by SNR-based gene ranking method. That is, the entire gene-list should be rearranged according to a specific ranking statistic and the original GSEA employs the SNR for gene ranking. The signal-to-noise ratio (SNR) for a gene i is given as follows.

Fig. 2. A typical graphical pattern of the SNR values for the gene-list ordered by SNR-based gene ranking

$$SNR(i) = \frac{\mu_A(i) - \mu_B(i)}{\sigma_A(i) + \sigma_B(i)} \tag{1}$$

Here the μ_A and μ_B denote the means of expression intensities of a gene i over sample group A and sample group B, respectively. The σ_A and σ_B denote the standard deviations of expression intensities of a gene i over sample group A and sample group B, respectively. When the SNR is used for gene ranking, the SNR values for the ordered gene-list typically have such a pattern as shown in Fig. 2. That is, the genes having higher SNR values (i.e. larger positive SNR values) are more highly ranked and are located at the front side of Fig. 2. The genes having lower SNR values (i.e. larger negative SNR values) are more lowly ranked and are located at the rear side of Fig. 2.

From eq. (1), we note that the positive SNR of a gene occurs when the gene has larger expression intensity in a sample group A than a sample group B, and the negative SNR of a gene occurs when the gene has smaller expression intensity in a sample group A than a sample group B.

With the entire gene-list ordered by this SNR ranking statistic and the genes included in a specific gene-set, the KS score is computed by employing the empirical cumulative distribution functions for P_{hit} and P_{miss} shown in eqs. (2).

$$P_{hit}(i) = \sum_{j=1}^{i} \frac{E(j)}{N_H}$$
$$P_{miss}(i) = \sum_{j}^{i} \frac{(1 - E(j))}{N_M} \tag{2}$$

$$ES = \max_{i=1,\cdots,N} \left| P_{hit}(i) - P_{miss}(i) \right| \tag{3}$$

Here P_{hit} is the empirical cumulative distribution function of which cumulative sum becomes 1 when the first i genes in the ordered gene-list completely match the genes included in a specific gene-set. On the other hand, P_{miss} is the one of which cumulative sum becomes 1 when there is no match between them. In eqs. (2), $E(j)$ is the ranking score (i.e. SNR) of the jth gene if the jth gene in the ordered gene-set is included in a gene-set (i.e. *hit*), or is 0 if the jth gene in the ordered gene-set is not included in a gene-set (i.e. *miss*). Also, N_H is the sum of ranking scores of the genes included in a gene-set, N_M is the sum of ranking scores of the genes not included in a gene-set, and N is the total number of genes in entire gene-list. Finally, the ES for a gene-set is obtained by taking the maximum absolute difference between P_{hit} and P_{miss} over all the genes, as in eq. (3). This means that the ES is determined at the maximum deviation point from zero in Fig. 3.

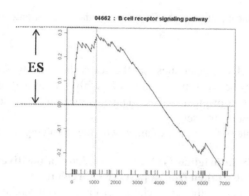

Fig. 3. The ES is determined at the maximum deviation from zero [1]

If the ES of a gene-set is chosen at the positive region, significant genes for the gene-set are taken from the left side of the maximum deviation point. That is, in this case, only the highly up-regulated genes in sample group A, compared with sample group B, are taken as significant ones. On the other hand, if the ES is chosen at the negative region, significant genes for the gene-set are taken from the right side of the maximum deviation point. That is, only the highly down-regulated genes in sample group A, compared with sample group B, are taken as significant ones. Because of this reason, both highly up-regulated genes and highly down-regulated genes cannot be chosen as significant ones at the same time for a specific gene-set, which makes it hard to reflect some situations incurred in biological pathways. So, we are interested in investigating a new gene ranking method for the gene-set enrichment analysis.

3.2 Fisher's Criterion Based Gene Ranking

As a ranking statistic for gene ranking, Fisher's criterion (FC) was used. For a gene i, the FC is given as below:

$$FC(i) = \frac{\left(\mu_A(i) - \mu_B(i)\right)^2}{\sigma_A(i)^2 + \sigma_B(i)^2}. \tag{4}$$

where μ_A, μ_B and σ_A, σ_B are the means and the standard deviations of gene expression intensities for sample groups A and B, respectively. When FC is used for gene

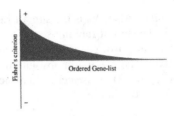

Fig. 4. A typical graphical pattern of the FC values for the gene-list ordered by FC-based gene ranking

ranking, there are such characteristics that the genes showing significant difference in expression intensities between two groups are highly ranked and located at the front side of Fig. 4, not as in original GSEA. Thus, we need a new selection strategy for determining the ES of a gene-set.

The selection strategy for ES is explained with the following three possible cases.

[CASE 1] When ES in original GSEA is taken from a positive SNR region and ES in FC-GSEA is also taken from a positive FC region
In this case, for a gene-set, the ES is chosen at the point of having maximum deviation from zero in both original GSEA and FC-GSEA. Significant genes are taken from the left side of the maximum deviation ES point.

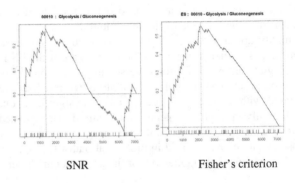

SNR   Fisher's criterion

Fig. 5. An example of ES located in a positive region for both original GSEA and FC-GSEA

[CASE 2] When ES in original GSEA is taken from a negative SNR region and ES in FC-GSEA is taken from a positive FC region
In this case, ES is chosen at the point of having maximum deviation from zero in both original GSEA and FC-GSEA, but in original GSEA, the actual value of ES should be taken as an absolute value of the maximum deviation ES and significant genes are identified from the right side of the maximum deviation ES point. On the other hand, note that in FC-GSEA, significant genes should be taken from the left side of the maximum deviation ES point.

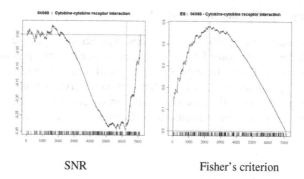

SNR   Fisher's criterion

Fig. 6. An example of ES located in a negative region for original GSEA but located in a positive region for FC-GSEA

[CASE 3] When ES in original GSEA is taken from a positive or negative SNR region and ES in FC-GSEA is taken from a negative FC region
This case reflects the situation that the maximum deviation ES point in FC-GSEA occurs in a negative region. Since the FC values are always positive, the negative quantity of ES means that there are no matching genes with a given gene-set. Thus, in such a case, we take the maximum deviation from zero in a positive region as ES in FC-GSEA for a gene-set, and significant genes are taken from the left side of the maximum deviation ES point. On the other hand, in original GSEA, significant genes are taken from the left or right side of the maximum deviation ES point, depending on whether the ES point is located either in a positive region or in a negative region.

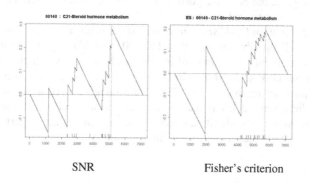

SNR   Fisher's criterion

Fig. 7. An example of ES located in a negative region for FC-GSEA

Thus, by using FC-based gene ranking for the gene-set enrichment analysis, it is possible to select both highly up-regulated genes and highly down-regulated genes at the same time as the significant genes for a gene-set. On the other hand, recall that the SNR-based gene ranking chooses either only the highly up-regulated genes or only the highly down-regulated genes as significant ones for a gene-set.

3.3 Estimating Significance in FC-GSEA

To estimate the statistical significance of an observed ES for a gene-set, it needs to be compared with the set of scores ES_{NULL} computed with randomly permutated samples with respect to sample labels. In original GSEA, a null distribution of ES is like Fig. 8. In Fig. 8, the left graph shows the case of locating an observed ES in a positive region while the right graph shows the case of locating an observed ES in a negative region. Also the bold-lined vertical bar indicates the location of an observed ES and the x axis is the permutation ES.

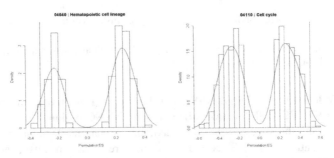

Fig. 8. An example of a null distribution of ES in original GSEA

In above cases, the estimation of significance for a gene-set is made with an observed ES whether it is in a positive region or in a negative region. If the observed ES is in a positive region, the positive portion of the distribution greater than the observed ES is used to estimate a nominal p-value for a gene-set. If the observed ES is in a negative region, the negative portion of the distribution less than the observed ES is used to estimate a nominal p-value.

On the other hand, in FC-GSEA, an observed ES is always located in a positive region. Thus, a null distribution of ES in FC-GSEA is like Fig. 9. Here the left (the right) graph corresponds to the case of having an observed ES in a positive (negative) region in original GSEA. As shown in Fig. 9, in either case, since an observed ES is always in a positive region for FC-GSEA, the positive portion of the distribution greater than the observed ES is used to estimate a nominal p-value for a gene-set. Thus, in order to identify significant gene-sets, one tailed test is performed in FC-GSEA while two-tailed test is performed in original GSEA.

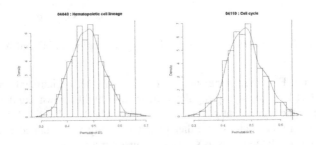

Fig. 9. An example of a null distribution of ES in FC-GSEA

3.4 Multiple Hypothesis Test in FC-GSEA

Once the ES is obtained for each gene-set, the ES should be normalized to account for the size of the gene-set, yielding a normalized ES (NES). The eq. (5) is the formula to calculate NES.

$$NES(S) = \frac{ES(S)}{\underset{\pi=1,\cdots,\Pi}{Mean}\big(ES_{\pi}(S)\big)} \tag{5}$$

Here $ES(S)$ is the enrichment score for a specific gene-set S and $ES_{\pi}(S)(\pi=1,\ldots,\Pi)$ is the ES for the permutation π of a specific gene-set S. Each permutation $\pi(\pi=1,\ldots,\Pi)$ is obtained by performing the random sampling of a gene-set S according to sample label and the total Πnumber of the permutated ESs are generated for a specific gene-set S. The number of permutation grows exponentially with the increasing number of samples and $\Pi=1000$ is generally used [1]. The significant gene-sets are identified by taking an appropriate number of gene-sets from the list of candidate gene-sets arranged by the normalized ES in a decreasing order. When FWER (Family-Wise Error Rate) test or FDR (False Discovery Rate) test is used to identify significant gene-sets, too small number of gene-sets can be chosen at times. In such cases, significant gene-sets can be identified with the normalized ES.

4 Experiments and Results

For our experiments, the Leukemia dataset [4] was used which includes 7129 human gene expression profiles of total 38 leukemia samples belonging to two classes, i.e. 27 acute lymphoblastic leukemia(ALL) samples and 11 acute myeloid leukemia(AML) samples. To identify significant pathways showing differential expression between ALL and AML classes, we first generated 167 candidate pathway gene-sets by taking the pathways each of which include at least five genes from KEGG pathway databases [11, 12]. Out of these pathways, the most significant 40 pathways were then found by using original GSEA and FC-GSEA in the same way as in [1], respectively. Also, for biological evaluation of the obtained pathways, a priori known leukemia-related pathways were manually collected with references to literature [15, 16, 17], Genetic Association Database, and KEGG pathway database and then used as the golden standards for biological verification.

4.1 Identification of Significant Genes

To identify significant genes for each of candidate pathways, we applied original-GSEA and FC-GSEA methods for the experimental dataset. As mentioned earlier, since the both have different selection strategy and expression difference metrics, the resulting genes which are differentially regulated ones between ALL and AML groups showed clear discrepancies by using original-GSEA and FC-GSEA, respectively. For example, significant genes identified from the gene-set of "B cell receptor signaling pathway" are as follows.

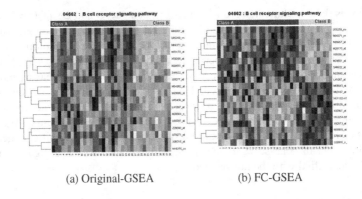

Fig. 10. Comparison of statistically significant genes identified by original-GSEA and FC-GSEA (bright grey boxes: results by original-GSEA, dark grey boxes: results by FC-GSEA)

(a) Original-GSEA   (b) FC-GSEA

Fig. 11. Heat maps of statistically significant genes identified by (a) original-GSEA and (b) FC-GSEA

As can be seen in Fig. 10, original GSEA that employs SNR-based gene ranking has the characteristics that the significant genes are selected by taking either genes whose SNR values are large in a positive region or genes whose SNR values are large in a negative region from the reference point. Consequently, the resulting significant genes are shown to have either only the highly up-regulated genes, as in Fig. 11 (a), or only the highly down-regulated genes in one group over the other. On the other hand, FC-GSEA that employs FC-based gene ranking has the characteristics that significant genes are chosen in such a way to take genes whose FC values are large in a positive region, i.e. genes having the relatively large expression difference between two groups. Thus, the resulting significant genes are shown, as in Fig. 11 (b), to have highly up-regulated genes and highly down-regulated genes together.

4.2 Finding Leukemia-Related Pathways

Specifically, from *literatures* [15, 16, 17], we first identified five pathways which are known to show the difference between AML and ALL types, i.e. *hsa04110* (cell cycle), *hsa04210* (apoptosis), *hsa04660* (T cell receptor signaling pathway), *hsa04662* (B cell receptor signaling pathway), and *hsa04640* (hematopoietic cell lineage). Two pathways including *hsa00480* (Glutathione metabolism) and *hsa00980* (Metabolism of xenobiotics by cytochrome P450) were obtained by taking the ones having p-value ≤0.05 by *Fisher's exact test*[19], out of the KEGG pathways including the AML-related or ALL-related genes identified from the Genetic Association Database

(GAD), which is the database provided by NIH that includes diseases-related genes and their detailed information.

Additionally, six pathways were identified by finding the KEGG pathways belonging to *Acute myeloid leukemia pathway*, which include *hsa04640*, *hsa04010* (MAPK signaling pathway), *hsa04630* (Jak-STAT signaling pathway), *hsa04210*, *hsa04150* (mTOR signaling pathway), and *hsa04110*.

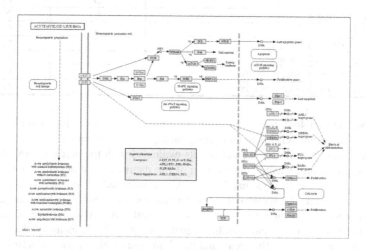

Fig. 12. The acute myeloid leukemia pathway appeared in KEGG pathway

Thus, by combining all of these pathways, we eventually collected 10 Leukemia-related pathways in total, as shown in Table 1, which were used as gold standards for the evaluation of our analysis results. Also, Table 2 shows the significant pathways on the Leukemia dataset each identified by original GSEA and FC-GSEA, respectively. As seen in this table, only the three pathways (hsa00480, hsa04110, hsa04640) were found with original-GSEA, On the other hand, by using FC-GSEA, eight of the ten pathways which include hsa00480, hsa00980, hsa04110, hsa04210, hsa04630, hsa04640,

Table 1. 10 Leukemia-related pathways collected by the authors

KEGG pathway names	Literatures	Fisher's exact test	Acute myeloid leukemia pathway
Glutathione metabolism		O	
Metabolism of xenobiotics by cytochrome P450		O	
MAPK signaling pathway			O
Cell cycle	O		O
mTOR signaling pathway			O
Apoptosis	O		O
Jak-STAT signaling pathway			O
Hematopoietic cell lineage	O		O
T cell receptor signaling pathway	O		
B cell receptor signaling pathway	O		

Table 2. Significant pathways identified by original-GSEA and FC-GSEA

KEGG pathway names	Original-GSEA	FC-GSEA
Glutathione metabolism	O	O
Metabolism of xenobiotics by cytochrome P450	X	O
MAPK signaling pathway	X	X
Cell cycle	O	O
mTOR signaling pathway	X	X
Apoptosis	X	O
Jak-STAT signaling pathway	X	O
Hematopoietic cell lineage	O	O
T cell receptor signaling pathway	X	O
B cell receptor signaling pathway	X	O

hsa04660 and hsa04642 were identified. As a result, this shows that FC-GSEA has its superiority to original-GSEA in identifying biologically significant pathways.

Table 3 shows our experimental results evaluated with the precision and the recall [20]. The precision indicates the fraction of the pathways that actually turn out to be positive out of the pathways that the GSEA has predicted to be significant pathways. On the other hand, the recall indicates the fraction of actually significant pathways correctly predicted by the GSEA out of the total pathways. Also, the precision and the recall can be summarized into another metric known as the F_1 measure. The formulas to calculate the precision, the recall and the F_1 measure [20] are shown below.

$$\text{Precision, } p = \frac{TP}{TP + FP}$$

$$\text{Recall, } r = \frac{TP}{TP + FN} \tag{6}$$

$$F_1 = \frac{2rp}{r + p} = \frac{2 \times TP}{2 \times TP + FP + FN}$$

Here TP is the number of true positive pathways, which is the number of the identified significant pathways by GSEA out of the collected 10 Leukemia-related pathways. FN is the number of false negative pathways, which is the number of the collected 10 Leukemia-related pathways not identified by GSEA. Also, FP is the number of false positive pathways, which is the number of the identified significant pathways by GSEA excluding the 10 Leukemia-related pathways.

Note that in the precision, the recall, and the F1 measure, the larger quantity indicates the better performance. According to Table 3, it is observed that FC-GSEA is also superior to original GSEA.

Table 3. Original-GSEA and FC-GSEA result's estimation

	Original-GSEA	FC-GSEA
Precision	0.3	0.8
Recall	0.075	0.2
F_1	0.12	0.32

5 Discussions

In this paper we investigated FC-GSEA method which employs Fisher's criterion for gene ranking in the gene set enrichment analysis and studied its applicability and usefulness via experiments on Golub's Leukemia datasets. As expected, our experiment results showed that the use of Fisher's criterion for gene ranking enables us to identify biologically significant pathways more extensively than original-GSEA approach employing SNR-based gene ranking, even if it is only one case. As future works, it seems to be worthwhile to perform further detailed and extensive analyses with FC-GSEA in a variety of contexts.

Acknowledgments. This work was supported by the Korea Science and Engineering Foundation (KOSEF) grant funded by the Korea government (MEST) (No. R01-2008-000-11089-0), and supported by the Korea Research Foundation Grant funded by the Korean Government (MOEHRD) (KRF-2008-331-D00558).

References

1. Subramanian, A., et al.: Gene set enrichment analysis: a knowledge-based approach for interpreting genome-wide expression profiles. Proc. Natl. Acad. Sci. USA 102, 15545–15550 (2005)
2. Zhang, A.: Advanced analysis of gene expression microarray data. World Scientific, Singapore (2006)
3. McLachlan, et al.: Analyzing Microarray Gene Expression Data. Wiley-Interscience/John Wiley & Sons (2004)
4. Taskesen, E.: Sub-typing of model organisms based on gene expression data. Bioinformatics Technical report, University of Delft (2006)
5. Monti, S., et al.: Molecular profiling of diffuse large B-cell lymphoma identifies robust subtypes including one characterized by host inflammatory response. Blood 1 105(5), 1851–1861 (2005)
6. Golub, T.R., et al.: Molecular classification of cancer: class discovery and class prediction by gene expression monitoring. Science 286, 531–537 (1999)
7. Bishop, C.: Neural Networks for Pattern Recognition. Oxford University Press, Oxford (1995)
8. Blum, A., et al.: Selection of relevant features and example in machine learning. Artificial intelligence 97, 245–271 (1997)
9. Bradley, P., et al.: Feature selection via mathematical programming. Technical report to appear in INFORMS Journal on computing (1998)
10. Tusher, V.G., et al.: Significance analysis of microarrays applied to the ionizing radiation response. Proc. Natl. Acad. Sci. 24 98(9), 5116–5121 (2001)
11. KEGG: Kyoto Encyclopedia of Genes and Genomes, http://www.genome.ad.jp/kegg/
12. Kanehisa, M., et al.: The KEGG databases at GenomeNet. Nucleic Acids Res. 30, 42–46 (2002)
13. Dinu, I., et al.: Improving GSEA for analysis of biologic pathways for differential gene expression across a binary phenotype. Collection of Biostatistics (2007)

14. Manoli, T., et al.: Group testing for Pathway analysis improves comparability of different microarray datasets. Bioinformatics 22(20), 2500–2506 (2006)
15. Kudsen, S.: Cancer Diagnostics with DNA Microarrays. John Wiley & Sons, Inc., Chichester (2006)
16. Potten, C., et al.: Apoptosis. Cambridge University Press, Cambridge (2005)
17. Weinberg, R.A.: The Biology of Cancer: Global surveys of Gene Expression Arrays. Garland Science (2006)
18. Becker, K.G., et al.: The genetic association database. Nature Genetics 36(5), 431–432 (2004)
19. Huang, D., et al.: Identifying the biologically relevant gene categories based on gene expression and biological data: an example on prostate cancer. Bioinformatics 15 23(12), 1503–1510 (2007)
20. Tan, P.N., et al.: Introduction to Data Mining. Addison Wesley, Reading (2006)
21. Dudoit, S., et al.: Multiple Hypothesis Testing in Microarray Experiments. Statistical Science 18, 71–103 (2003)

A Comparison between Quaternion Euclidean Product Distance and Cauchy Schwartz Inequality Distance

Di Liu, Dong-mei Sun, and Zheng-ding Qiu

Institute of Information Science, Beijing Jiaotong University, Beijing, China, 100044
liud8310@gmail.com

Abstract. This paper proposes a comparison in handmetrics between Quaternion Euclidean product distance (QEPD) and Cauchy-Schwartz inequality distance (CSID), where "handmetrics" refers to biometrics on palmprint or finger texture. Previously, we proposed QEPD [1,2] and CSID [11] these two 2D wavelet decomposition based distances for palmprint authentication and face verification respectively. All two distances could be constructed by quaternion which was introduced for reasonable feature representation of physical significance, i.e. 4-feature parallel fusion. Simultaneously, such quaternion representation enables to avoid incompatibleness of multi-feature dimensionality space for fusion. However, a comparison between two distances is seldom discussed before. Therefore, we give a comparison on experimental aspects for providing a conclusion which algorithm is better. From the result, we can conclude the performance of QEPD is better than CSID and finger texture is a better discriminative biometric than palmprint for QEPD or CSID.

1   Introduction

Biometrical information fusion community is flooded with a large number of biometric fusion approaches. Traditionally, in field of the multi-feature biometric fusion, it is well known that these schemes are categorized into four major levels, namely (a) sensor level, i.e. data level, (b) feature level, (c) matching score level, and (d) decision-making level [7]. Recently, it raises a heated research on feature level, i.e. Feature Fusion Level (FLF) in the community of multi-feature biometric recognition, because of (a) richer information which can be acquired from raw data than other level fusion and (b) good performance without independence assumption rather than decision-making level. And FLF classifies into four aspects based on method fusion, namely serial fusion, parallel fusion, weight fusion, and kernel fusion. Traditionally, serial fusion is a most common used fusion method, yet it has two evident drawbacks: (1) the dimensionality of feature vector after concatenation increases dramatically so that recognition speed goes down sharply. (2) as to small sample size, e.g. face recognition, this method usually gives rise to singularity which is an obstacle of feature extraction of Fisher criterion. To avoid the problems, [10] proposed parallel fusion with a

T.-h. Kim et al. (Eds.): FGCN 2008 Workshops and Symposia, CCIS 28, pp. 129–137, 2009.

complex feature vector form that aims at obtaining a higher recognition effect and a faster speed than serial fusion. According to the advantage above, a wavelet decomposition feature based quaternion fusion with QEPD [1,2] for palmprint or finger texture authentication was proposed. Also, another one with CSID [11] migrated from face verification is implemented for palmprint or finger in this paper. We propose a reasonable interpretation for the mathematical significance of 4-feature parallel fusion by introducing quaternion prior to a discrimination by such two distances respectively. Then we conduct a comparison on experimental aspects for providing a conclusion which algorithm is better. The rest of paper is organized as follows: section 2 introduces QEPD and CSID in turn. The next section we propose our experimental comparison between such two distances for handmetrics verification in details. After that, section 4 provides a conclusion.

2   QEPD and CSID

2.1   Quaternion Euclidean Product Distance (QEPD)

We here make QEPD available as a discriminant distance. The relationship between QEP and quaternion modulus was discussed in [1,2]. Thus for an arbitrary pixel corresponding to 4 separable wavelets decomposition sub-image, consider two quaternions, which the former is from this pixel as the template, $P = a + bi + cj + dk$ and the latter is from the tester, in which $Q = t + xi + yj + zk$. Ideally, if $P = Q$ such that $\overline{P}P = |P|^2$, where is a particular case of Quaternion Euclidean product. We can estimate the difference between the template quaternion matrix and the tester one by QEPD $D(\overline{P}P, |\overline{P}Q|)$ as a discriminant distance. In which $|\overline{P}Q|$ is an entry of the matrix with the modulus of QEP, and the operator D is certain kind of distance, e.g. L_1 norm, L_∞ norm distance, Euclidean distance etc. Such two matrices have the same size as the subimage above. The reason use the modulus is that the multiplication of two different quaternions is a quaternion so that it is impossible to compare directly with $\overline{P}P$. Because $\overline{P}Q$ is not a scalar as follows.

$$
\begin{aligned}
\overline{P}Q &= (a - bi - cj - dk)(t + xi + yj + zk) \\
&= (at + bx + cy + dz) + (ax - bt - cz + dy)i \\
&\quad +(ay + bz - ct - dx)j + (az - by + cx - dt)k
\end{aligned} \tag{1}
$$

Therefore, the similarity between template and tester QEPD $D(\overline{P}P, |\overline{P}Q|)$ is obtained by Euclidean distance between the matrix of the absolute value of the equation (1) and that of template's modulus square $|P|^2$.

2.2   Cauchy-Schwartz Inequality Distance (CSID)

(1) Autocorrelation for signal processing. Autocorrelation is used frequently in signal processing for analyzing functions or series of values, such as time domain signals [6]. We here discuss the case of discrete signal sequence, defining

$$
R_{xx}(j) = \sum_n x_n \overline{x}_{n-j} \tag{2}
$$

as autocorrelation for x_n. This term has many properties, e.g.symmetry, $R_{xx}(-\tau)$ = $R_{xx}(\tau)$ when $x(n)$ is a real sequence, and $R_{xx}(-\tau) = R_{xx}^*(\tau)$ when $x(n)$ is a complex one. In addition, a wavelet feature based quaternion representation for parallel fusion is viewed as a real signal sequence.

(2) Cauchy-Schwartz inequality for autocorrelation. According to the section above, another significant property of autocorrelation is Cauchy-Schwartz inequality [6]. Given two arbitrary vectors a, b their inner product space is depict as $|\langle a, b \rangle|^2 \le \langle a, a \rangle \cdot \langle b, b \rangle$ and the inequality is expressed as $|\langle a, b \rangle| \le \|a\| \cdot \|b\|$, where $\langle \cdot \rangle$ is inner product, and $\|\cdot\|$ denotes norm of the vectors. Such consequence is utilized for property of autocorrelation as follows

$$|R_{xx}(\tau)| \le R_{xx}(0) \tag{3}$$

The equation (2) is used for construction of CSID distance, by which discriminate genuine and impostor.

(3) Cauchy-Schwartz inequality distance. Consider two quaternions, which the former is $P = a + bi + cj + dk$ and the latter from the tester, in which $Q = t + xi + yj + zk$. If template and tester one belong to the same person, P will be more similar with Q than that from different persons, i.e. $|P - Q|$, modulus of $P - Q$, is smaller than that from different persons. Thus we obtain

$$\lim_{Q \to P} |P - Q| = 0 \tag{4}$$

Thus we view P and Q as real signal sequences in sense of discrete-time signals, i.e. transform quaternion P and Q into forms of $\{x(p)|a, b, c, d\}$ and $\{x(q)|t, x, y, z\}$ in which $x(p)$ and $x(q)$ are discrete sequences at time p and q respectively. According to the autocorrelation discussed above, the equation (3) can be rewritten as

$$R_{xx}(\tau) = \sum_n x_p x_q \tag{5}$$

Where $p = q - \tau$, time q can be viewed as a time delay to p. Now evolve the equation (3)

$$R_{xx}(0) - |R_{xx}(\tau)| \ge 0 \tag{6}$$

Replace (5) with (6), we obtain

$$D_{pixel} = \sum_n x_p^2 - \left| \sum_n x_p x_q \right| \ge 0 \tag{7}$$

It is easily found that the equation above is larger than 0. To this end, it is accommodate to set this as the distance of the pixel for 4 sub-images with a reasonable physical significance. For all pixels of such sub-images, we define Cauchy-Schwartz inequality distance

$$D_{CSID} = \sum_{i=1}^{k} D_{pixel} = \sum_{i=1}^{k} \left\{ \sum_n x_p^2 - \left| \sum_n x_p x_q \right| \right\} \ge 0 \tag{8}$$

Where k is number of pixel in such sub-images, n is the component number of discrete signal sequence, e.g. $k = 4$ because of the quaternion. Notice that D_{CSID}, the sum of D_{pixel}, evidently has a characteristic of $D_{CSID} \geq 0$.

3   Comparison by Experiment

3.1   Database

BJTU-HA biometric database, an inherited collection work by Institute of Information Science, Beijing Jiaotong University, is utilized for our palmprint and middle finger texture verification experiment. It contains totally 1,500 samples from 98 person's palmprint and middle finger with different illumination conditions. In the experiment, we use a subset of the database with 10 samples from first 40 persons, totally 400 images, as our matching set.

3.2   QEPD Computation

After quaternion fusion, we employ QEPD for matching. Suppose A and B as two quaternion vectors of palmprint or finger texture feature, $A = \{a + bi + cj + dk\}$, where a, b, c, d are wavelets decomposition coefficients respectively. The same is true of $B = \{t + xi + yj + zk\}$. According to this distance, two matching score tables (Table 1 and Table 2) are listed from the following palmprint (figure 1) and finger texture ROI images (figure 2) respectively as matching examples:

Sample matching in the database is proposed for investigate the performance of our scheme. That is, each sample is matched with other samples in the subset

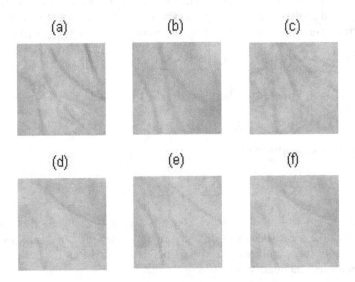

Fig. 1. Palmprint ROI sample for matching. In the experiment, each ROI sample with a 128×128 dimensionality has been used.

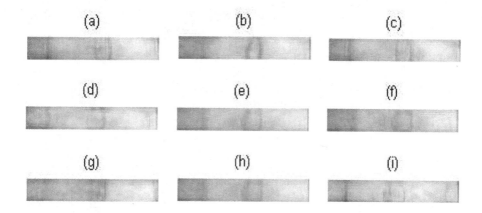

Fig. 2. Finger texture ROI sample for matching. In the experiment, each ROI sample with a 12×72 dimensionality has been used.

Table 1. Palmprint QEPD computation among figure 1

Fig	1(b)	1(c)	1(d)	1(e)	1(f)
1(a)	0.273	0.245	0.264	0.299	0.236
1(b)		0.270	0.347	0.382	0.322
1(c)			0.281	0.256	0.271
1(d)				0.221	0.060
1(e)					0.239

Table 2. Finger texture QEPD computation among figure 2

Fig	2(b)	2(c)	2(d)	2(e)	2(f)	2(g)	2(h)	2(i)
2(a)	0.245	0.149	0.382	0.202	0.265	0.436	0.242	0.397
2(b)		0.202	0.345	0.279	0.314	0.390	0.260	0.423
2(c)			0.292	0.268	0.343	0.422	0.300	0.384
2(d)				0.408	0.509	0.431	0.457	0.386
2(e)					0.178	0.419	0.160	0.417
2(f)						0.362	0.072	0.496
2(g)							0.331	0.475
2(h)								0.477

we defined. The matching between the samples from the same person is defined as inner-class matching; otherwise, it is classified as outer-class matching. Totally, about 160,000 (400×400) matching have been performed, where 3,600 matching are inner-class matching. Figure 3(a) and (b) show the distribution of inner-class and outer-class discriminant distance of palmprint and middle finger texture respectively. From them, it is easily found that peak of palmprint inner-class QEPD is at about matching score 0.1, and 0.12 by finger texture, where the

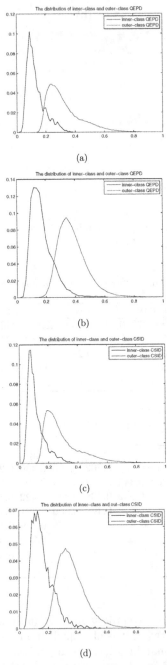

Fig. 3. (a) The distribution of palmprint inner-class and outer-class QEPD (b) the distribution of middle finger texture inner-class and outer-class QEPD (c) the distribution of palmprint inner-class and outer-class CSID and (d) the distribution of middle finger texture inner-class and outer-class CSID

discriminant distances are normalized, i.e. each QEPD divides by the maximum of QEPD. Likewise the peak of palmprint outer-class approaches 0.24 and 0.34 by finger. The two peaks can be separated widely, hence scheme can effectively discriminate these handmetric images.

3.3 CSID Computation

According to CSID discussed previous section, discriminant distance tables corresponding to figure 1 and 2 are shown in table 3 and 4.

Totally, about 160,000 (400×400) matching of palmprint and finger texture have been performed respectively, where 120,000 matching are outer-class CSID for each. Figure 3(c) and (d) show the distribution of inner-class and outer-class CSID. From them, peak of inner-class CSID of palmprint and finger texture are at about 0.09 and 0.14. And the peak of outer-class approach 0.21 and 0.32. The two peaks can be separated widely, so the scheme discriminate them like QEPD.

3.4 Comparison of Result

Firstly, we make a comparison among four distribution graphes. From the following table, it is found that as to distance between inner-class and outer-class peak, i.e. difference in the table 5, finger QEPD is widen than palmprint's, and the same is true that CSID. This means QEPD may discriminant effectively than CSID. Interestingly, regardless of QEPD or CSID, finger's is large than that of palmprint. Here we provide an inference that this finger texture proposed is an effective biometric. Then we will use ROC performance for testifying this inference.

Table 3. Palmprint CSID computation among figure 1

Fig	1(b)	1(c)	1(d)	1(e)	1(f)
1(a)	0.523	0.445	0.541	0.592	0.479
1(b)		0.518	0.692	0.778	0.640
1(c)			0.564	0.504	0.534
1(d)				0.416	0.113
1(e)					0.446

Table 4. Finger texture CSID computation among figure 2

Fig	2(b)	2(c)	2(d)	2(e)	2(f)	2(g)	2(h)	2(i)
2(a)	0.224	0.147	0.386	0.196	0.258	0.408	0.227	0.393
2(b)		0.173	0.318	0.274	0.300	0.383	0.252	0.416
2(c)			0.289	0.259	0.349	0.409	0.298	0.375
2(d)				0.406	0.532	0.419	0.475	0.387
2(e)					0.186	0.394	0.151	0.399
2(f)						0.354	0.072	0.496
2(g)							0.317	0.451
2(h)								0.475

Table 5. Comparison of four distribution graphes

	palmprint QEPD	finger QEPD	palmprint CSID	finger CSID
Peak of inner-class	0.10	0.12	0.09	0.14
Peak of outer-class	0.24	0.34	0.21	0.32
Difference	0.14	0.22	0.12	0.18

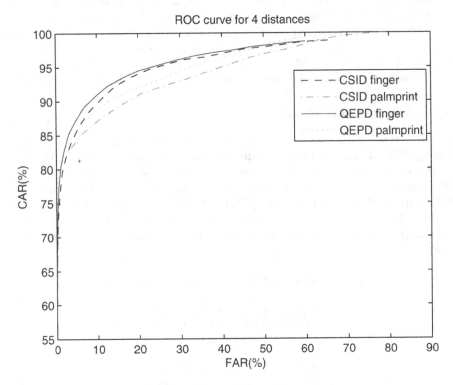

Fig. 4. ROC curve for 4 distances

The performance of a verification approach is measured by False Accept Rate (FAR) and Correct Accept Rate (CAR). The ROC curve for all four distances is shown in Figure 4. From the ROC curves for 4 distances, the performance of finger QEPD, finger CSID, palmprint QEPD, and palmprint CISD size down in turn. Therefore, we can safely conclude that QEPD outperforms CSID and finger texture is a better discriminative biometrics than traditional biometric palmprint for QEPD or CSID. This ROC curve has proven the inference above.

4   Conclusion

This paper mainly proposes a comparison by two different discriminate distance, Quaternion Euclidean Product Distance (QEPD) and Cauchy-Schwartz Inequality Distance (CSID), in order to solve space incompatibleness and curse of dimensionality in a non-subspace means. Here we call palmprint or finger texture biometric traits as "handmetric". From the experiments, we can safely provide two conclusions: (1) the algorithm of QEPD is better than that of CSID, (2) finger texture, this novel biometric is a better discriminative than traditional biometric palmprint for QEPD or CSID.

Acknowledgements

This work is fully supported by National Science Foundation of China under Grant No. 60773015 and Research Fund for the Doctoral Program of Higher Education No. 20060004007.

References

1. Liu, D., Sun, D.M., Qiu, Z.D.: Wavelet decomposition 4-feature parallel fusion by quaternion euclidean product distance matching score for palmprint verification. In: 9th International Conference on Signal Processing, pp. 2104–2017 (2008)
2. Liu, D., Qiu, Z.D., Sun, D.M.: 2D wavelet decomposition feature parallel fusion by quaternion euclidean product distance matching score for middle finger texture verification. In: 2008 International Conference on Bio-Science and Bio-Technology, December 13-15 (2008) (in press)
3. Ross, A., Jain, A.: Information fusion in biometrics. Pattern Recognition Letters 24(13), 2115–2125 (2003)
4. Lang, F., Zhou, J., Bin, Y., Song, E., Zhong, F.: Quaternion based information parallel fusion and its application in color face detection. In: 9th International Conference on Signal Processing, vol. 3, pp. 16–20 (2006)
5. Wen, L.: Quaternion Matrix, pp. 96–100. National University of Defense Technology Press, ChangSha (2002)
6. Proakis, J.G., Manolakis, D.G.: Digital Signal Processing. Pearson Education Press, London (2007)
7. Jain, A., Nandakumar, K., Ross, A.: Score normalization in multimodal biometric systems. Pattern Recognition 38(12), 2270–2285 (2005)
8. Sanderson, C., Paliwal, K.K.: Identity verification using speech and face information. Digital Signal Processing 14(5), 449–480 (2004)
9. Jain, A.K., Ross, A.: Multibiometric systems. Communication of the ACM 47(1), 34–40 (2004)
10. Yang, J., Zhang, D., Lu, J.: Feature fusion: parallel strategy vs. serial strategy. Pattern Recognition 36(6), 1369–1381 (2003)
11. Liu, D., Qiu, Z.D., Sun, D.M.: SIFT Feature-Based Face Verification by Bag-of-words Vector Quantization. Institute of Electronics, Information and Communication Engineering, transactions on information and systems (submitted)

Wavelet Decomposition Feature Parallel Fusion by Quaternion Euclidean Product Distance Matching Score for Finger Texture Verification

Di Liu Zheng-ding Qiu and Dong-mei Sun

Institute of Information Science, Beijing Jiaotong University, Beijing, China, 100044
liud8310@gmail.com

Abstract. Parallel fusion is a promising fusion method in field of feature level fusion. Unlike conventional 2-feature parallel fusion with complex representation, it is tough to represent 4-feature parallel fusion due to shortage of mathematical significance. To solve this problem, this paper proposes a reasonable interpretation by introducing quaternion. Initially, this paper defines a novel ROI extraction method for obtaining more information from middle finger images. After parallel fusion whose features are extracted by 2D wavelets decomposition coefficients from the same pixel corresponding to 4 separate sub-images, Quaternion Euclidean Product Distance (QEPD), a distance between modulus square of template quaternion and modulus of tester Quaternion Euclidean Product (QEP), as matching score, is performed. The scores discriminate between genuine and impostor of finger texture by threshold effectively. Finally, the experimental result gains a reasonable recognition rate but at a fast speed. Through a comparison with palmprint QEPD, the recognition performance of this finger texture QEPD outperforms than palmprint.

1 Introduction

Nowadays information fusion community is flooded with a large number of biometric fusion approaches. Obviously, information fusion is an information process utilized to computer technique to automatically analyze information obtained by sensors serially based on specific criterions, for accomplishing the need of decision and estimation [1]. At present, we study fusion algorithm on certain specific context, such as biometric fusion, a novel research topic in information fusion. Traditionally, in field of the biometric fusion, it is well known that these schemes are categorized into four major levels, namely (a) sensor level, i.e. data level, (b) feature level, (c) matching score level, and (d) decision-making level [2,3,4].

(a) **Fusion at data level:** It is a low-level fusion that data extracted from each sensor of an arbitrary object are concatenated into a higher dimensional raw data matrix, e.g. a low unified quality image fusion by multiple channel information, and then it is used to identification or verification. At the level, many build-up

T.-h. Kim et al. (Eds.): FGCN 2008 Workshops and Symposia, CCIS 28, pp. 138–149, 2009.

algorithms are proposed and the ideas are used in transform-domain of original image [5].

(b) **Fusion at feature level:** This intermediate-level fusion computes feature data from each sensor into a feature vector, for instance, each feature vector from sensor can be concatenate into a vector. It gains a better recognition performance in theory than other fusion levels due to their storage of original information from data. But it usually raises problem of space incompatibleness, or curse of dimensionality.

(c) **Fusion at matching-score level:** Matching scores are provided by system indicate proximity of the feature vector with the template vector for personal identification or authentication [6].

(d) **Fusion at decision-making level:** This high-level fusion makes a fusion after matching of each trait. That is, it combines the classification result of each feature vector, i.e. accept or reject [6]. As a result, the fusion has a high robustness and low algorithm complexity.

Recently, it raises a heated research on feature level, i.e. Feature Fusion Level (FLF) in the community of multimodal biometric recognition, because of (a) richer information which can be acquired from raw data than other level fusion and (b) good performance without an independence assumption rather than decision-making level. And FLF classifies into four aspects based on fusion method, namely serial fusion, parallel fusion, weight fusion, and kernel fusion. Traditionally, serial fusion is a most common used fusion method, yet it has two evident drawbacks: (1) the dimensionality of feature vector after concatenation increases dramatically so that recognition speed goes down sharply. (2) as to Small Sample Size (SSS), e.g. face recognition, this method usually gives rise to singularity of the within class scatter matrix with a high dimensionality of concatenated feature matrix, which is an obstacle of feature extraction of Fisher criterion. To solve such problems, [7,8] proposed parallel fusion with a complex feature vector form that aims at obtaining a higher recognition effect and a faster speed than serial fusion. According to the advantage above, we propose an approach of 2D wavelet decomposition feature parallel fusion in terms of a characteristic of QEP for finger texture authentication. We propose a new finger texture ROI as our extraction ROI with more information for wavelets feature acquisition but at a low dimensionality. Then partial wavelet coefficients by level 2 wavelet decomposition are obtained as fusion features so as to avoid redundant feature data like all the coefficients used. After that, we use QEPD, a distance between modulus square of template quaternion and modulus of tester QEP as matching score for finger texture verification. For proving the effectiveness of experiment of finger texture QEPD, a comparison with palmprint QEPD is conducted. The final recognition result with 93.36% is higher than the same means with 92.87% by palmprint. In addition, the dimensionality of original image sample, composed quaternion matrix the system processes for finger texture, are lower than palmprint. Therefore, we can conclude such finger texture is a better biometric than palmprint for QEPD. Also, it is a fast time for the entire processing, through checking time cost of each procedure of finger texture QEPD.

The rest of paper is organized as follows: section 2 introduces basic issues of quaternion, concept of wavelet decomposition and parallel fusion. Then we propose our scheme for finger texture verification in the section 3. After that, section 4 shows that experimental results by the approach. Finally, we perform a conclusion.

2   Related Work

2.1   Quaternion

The quaternion is a non-commutative extension of complex numbers, which first described by the Irish mathematician Sir William Rowan Hamilton in 1843 and applied to mechanics in three-dimensional space [5,9]. Quaternion has one real part and three imaginary parts, not only one like complex number. Let $P = \{a + bi + cj + dk | a, b, c, d \in R\}$, where a, b, c, d are real numbers, and i, j, k are imaginary identities, respectively. According to the definition, quaternion has basic properties of

$$
\begin{aligned}
i^2 = j^2 = k^2 = ijk = -1, \\
ij = k, ji = -k, jk = -i, \\
kj = -i, ki = j, ik = -j
\end{aligned}
\tag{1}
$$

Also, quaternion has a lot of similarities as complex number, e.g.
Conjugate

$$
\overline{P} = a - bi - cj - dk
\tag{2}
$$

Modulus

$$
|P| = P\overline{P} = \sqrt{a^2 + b^2 + c^2 + d^2} \geq 0
\tag{3}
$$

$Q = \{t + xi + yj + zk | t, x, y, z \in R\}$ is another quaternion, and we discuss quaternion arithmetic operations in the following:
Equality

$$
P = Q \Leftrightarrow a = t, b = x, c = y, d = z
\tag{4}
$$

Addition

$$
P + Q = (a + t) + (b + x)i + (c + y)j + (d + z)k
\tag{5}
$$

Multiplication

$$
\begin{aligned}
P \times Q = (at - bx - cy - dz) + (bt + ax + cz - dy)i \\
+ (ct + ay + dx - bz)j + (dt + az + by - cx)k
\end{aligned}
\tag{6}
$$

Inner Product

$$
< P, Q >= at + bx + cy + dz
\tag{7}
$$

And Quaternion Euclidean Product (QEP) is defined as:

$$
QEP = \overline{P}Q
\tag{8}
$$

Consider that when $P = Q$, QEP is equal to $\overline{P}P = |P|^2$, i.e. the square of modulus.

2.2   The Extraction of 2-D Wavelet Decomposition Coefficients

Wavelets transform booms as a signal processing approach of transform domain in field of computer vision and image processing. With regard to wavelet, an extension of traditional Fourier transforms, its multi-resolution analysis has a good time-frequency characteristic so that it is feasible to process stationary signals, e.g. finger texture image. It is worth to say that wavelet coefficients can be considered as features in parallel fusion schemes. Concretely, a finger texture image is decomposed into four sub-images by Wavelets decomposition (Fig.1), which are approximation coefficients, horizontal coefficients, vertical coefficients, and diagonal detailed coefficients. The fig.1 proposes a 2D Wavelet Decomposition at level 2, and there are 4 sub-images with a scale of 1/4 on the left top corner. Each pixel of sub-images, i.e. coefficients with the scale of 1/4, are utilized for feature selection of parallel fusion because of its uniqueness, so as to perform biometric verification.

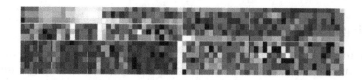

Fig. 1. 2D wavelet decomposition at level 2 for finger texture image

2.3   Parallel Fusion

Suppose that A and B in which are respectively feature vectors extracted from two traits, e.g. face, fingerprint etc. The parallel fusion is in the form of $c^l = \{a_1^l + ib_1^l, a_2^l + ib_2^l, \cdots\}$ where c^l denotes as a complex vector, and $i^2 = -1$. Provided that $m > n$, where m, n are subscripts of a_m^l and b_n^l, i.e. the numbers of samples, then set 0 as $b_{n+1}^l, b_{n+2}^l, \cdots, b_m^l$, i.e. $c^l = \{a_1^l + ib_1^l, a_2^l + ib_2^l, \cdots a_{n+1}^l, a_{n+2}^l \cdots a_m^l\}$; vice versa. According to quaternion [5,9], we make use of four wavelet decomposition coefficients i.e. approximation coefficient, horizontal coefficient, vertical coefficient, and diagonal detailed coefficient as parallel features. That is to say, a quaternion is constructed as $W = w_1 + w_2 i + w_3 j + w_4 k$, where w_1, w_2, w_3, w_4 are such four coefficients separately.

3   Our Algorithm

Basically, proposed algorithm workflow is illustrated in the fig. 2. In the phase of preprocessing, our ROI extraction method is proposed. Since feature extraction, 4 wavelet decomposition coefficients have been obtained by 2 level wavelets decomposition for composing a quaternion as the parallel fusion. Then matching module proposes matching between template quaternion and test one from which can gain a matching score. This score can be discriminated by the specific threshold as the decision.

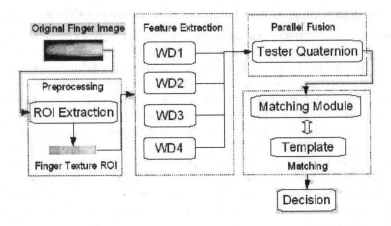

Fig. 2. 2D wavelet decomposition feature parallel fusion workflow

3.1   Finger Texture ROI Extraction

In order to obtain more information from finger texture for wavelet decomposition feature, we define a 12×72 window at the preprocessed finger image, as our feature ROI (Fig.3(b)). Here the ROI include most area about finger except fingerprint.

Fig. 3(b) ROI which includes most finger texture feature without fingerprint, is extracted by the window from the original image with a size of 30×90. The reason we choose the ROI size is that (1) more finger texture information for wavelets decomposition result in a better recognition performance and (2) the dimensionality of ROI proposed is suitable for computation of 2D wavelets decomposition.

3.2   Wavelet Decomposition Feature Parallel Fusion

In the algorithm, we obtain partial wavelets coefficients as our features. That is, each quaternion is composed of 4 sub-image coefficients as feature of parallel fusion, i.e. a group of wavelets coefficients of a pixel stand for a quaternion, thus a group of quaternions were likely to be acquired for an image, i.e. $\{P|P_n = c_1^n + c_2^n i + c_3^n j + c_4^n k\}, n = 1, 2, \cdots$ where n was sequence of pixel of sub-image by wavelet decomposition. Notice that traditional complex vector parallel fusion usually faces a common problem, incompatibleness of dimensionality of two features. As mentioned in the section 2.3, it is often case that two

Fig. 3. (a) Original image (b) ROI windows

features of the same sample have not identical number, e.g. $m > n$. If assign 0 as $b^l_{n+1}, b^l_{n+2}, \cdots, b^l_m$, this method will be lack of physical significance. Here our algorithms enable to avoid this problem because each sub-image pixel has four coefficients for fusion at the same time.

3.3 Quaternion Euclidean Product Distance

As mentioned section above, we here make QEPD available as matching score. The relationship between QEP and quaternion modulus has been discussed in the section 2.1, thus consider two quaternions, for an arbitrary pixel corresponding to 4 separable wavelets decomposition coefficients sub-image, which the former is from this pixel as the template, $P = a + bi + cj + dk$ and the latter is from the tester, in which $Q = t + xi + yj + zk$. Ideally, if $P = Q$ such that $\overline{P}P = |P|^2$, where is a particular case of QEP. We can estimate the difference between the template quaternion matrix and the tester one by QEPD $D(\overline{P}P, |\overline{P}Q|)$ as a discriminant distance. Notice that the template quaternion matrix and tester matrix refer to a matrix stores the value of $\overline{P}P$ corresponding to each pixel of the subimage and a matrix with all of $|\overline{P}Q|$ corresponding to these pixels respectively. Such two matrices have the same size as the subimage above. In which $|\overline{P}Q|$ is the modulus of Quaternion Euclidean product, and the operator D is certain kind of distance, e.g. L_1 norm, L_∞ norm distance, Euclidean distance etc. In our scheme, Euclidean distance is chosen as the operator D. The reason use the modulus of $\overline{P}Q$ is that the multiplication of two different quaternions is a quaternion (equation (9)), combined (3) and (6), so that it is impossible to compare directly with $\overline{P}P$. From the equation, $\overline{P}Q$ is not a scalar.

$$\begin{aligned}
\overline{P}Q &= (a - bi - cj - dk)(t + xi + yj + zk) \\
&= (at + bx + cy + dz) + (ax - bt - cz + dy)i \\
&\quad + (ay + bz - ct - dx)j + (az - by + cx - dt)k
\end{aligned} \qquad (9)$$

Therefore, the similarity between template and tester QEPD $D(\overline{P}P, |\overline{P}Q|)$ is obtained by Euclidean distance between the matrix of the absolute value of the equation (9) and that of template's modulus square $|P|^2$.

3.4 QEPD Matching

We make 2-D wavelets decomposition feature as parallel fusion discussed in the subsection 3.2. After fusion, we employ QEPD for matching. Suppose A and B as two finger texture feature quaternion vectors for each pixel of the subimage, $A = \{a + bi + cj + dk\}$ where a, b, c, d are wavelets decomposition coefficients respectively. The same is true like $B = \{t + xi + yj + zk\}$. According to QEPD $D(\overline{P}P, |\overline{P}Q|)$, a matching score table (Table 1) is listed from the following Finger texture ROI image (Fig.4):

From the table 1, QEPD matching scores are calculated by each two samples. The most intra-class scores of this QEPD matching score estimated in our experiment, usually span from 0 to 0.32, and majority of inter-class scores from

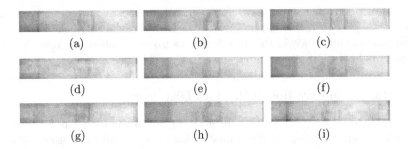

Fig. 4. Finger texture ROI for matching

Table 1. QEPD matching scores among the Fig.4

Fig	4(b)	4(c)	4(d)	4(e)	4(f)	4(g)	4(h)	4(i)
4(a)	0.2449	0.1492	0.3824	0.2015	0.2653	0.4358	0.2419	0.3973
4(b)		0.2021	0.3447	0.2786	0.3139	0.3898	0.2603	0.4234
4(c)			0.2921	0.2674	0.3427	0.4215	0.3001	0.3836
4(d)				0.4082	0.5089	0.4306	0.4570	0.3864
4(e)					0.1777	0.4189	0.1597	0.4165
4(f)						0.3618	0.0722	0.4955
4(g)							0.3314	0.4751
4(h)								0.4771

0.25 to 0.64. For example, fig.4(f) and fig.4(h) belong to the same class because of a low score of 0.0722. Conversely, fig.4(d) is a impostor to fig.4(f) with a high score of 0.5089. We will discuss the distribution of genuine and impostor in the subsection 4.2.

4 Experiment

4.1 Finger Database

BJTU-FI biometric database, an inherited collection work by Institute of Information Science, Beijing Jiaotong University, is utilized for our finger texture verification experiment. It contains totally 1,500 samples from 98 person's fingers with different illumination conditions. In order to avoid a high dimensional computation, we acquire middle finger ROI like in the fig.3(b) by our window. In the test section, we use a subset of the database with 10 samples from each person, totally 360 images, as our matching set.

4.2 Finger Matching

Sample matching in the database is proposed for investigating the performance of our scheme. That is, each sample is matched with others samples in the subset

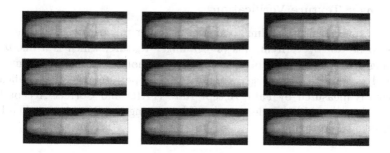

Fig. 5. Typical middle fingers samples in the BJTU-FI database

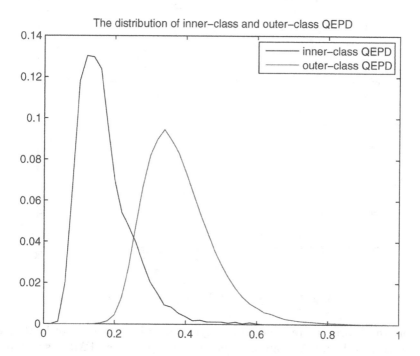

Fig. 6. The distribution of genuine and impostor finger texture QEPD matching score

we defined. The matching between the template and samples from the same person is defined as genuine matching; otherwise, it is classified as impostor matching. Totally, about 129,600 (360×360) matching have been performed, where 3,600 matching are genuine matching. Fig.6 shows the distribution of genuine and impostor matching scores. From the fig.6, it is easily found that peak of genuine is at about matching score 0.12, where the matching score is obtained by normalization of QEPD between template sample and tester one. Likewise, the peak of imposter is approached 0.36. The two peaks can be separated widely, hence the scheme effectively discriminate finger texture images.

4.3   Finger Texture Verification

Finger texture verification aims to answer 'whether this person is who he or she claims to be' [6] by verifying his or her finger texture, which is a one-to-one matching. If the matching score exceeds the threshold by proposed scheme, it is accepted as genuine, otherwise, as impostor. The performance of a verification approach is measured by False Accept Rate (FAR) and Correct Accept Rate (CAR). They should be made a trade-off by ploting the ROC curve. The ROC curve is shown on the fig.7.

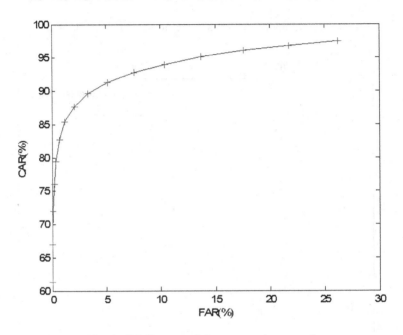

Fig. 7. ROC curve of the proposed approach

From the figure, the recognition rate goes up with an increase of FAR. The system keeps a high CAR rate i.e. from 80% to 90%, when FAR is small, e.g. less than 5%. It achieved 96% of recognition rate since FAR is at 25%.

SVM algorithms have been used as our classifier in this experimental section. Training set and testing set are totaly separated. In SVM, we assign 10 positive targets (inner-class) and 10×35 negative targets (outer-class) as training set for each person (class), so training set for 40 classes is $(10 + 350) \times 40 = 14,400$. Notice that the outer-class targets are chosen by each 10 samples from other 35 outer classes respectively. Finally, the recognition rate approaches 93.36% accuracy.

4.4   Speed

The experiment is implemented by Matlab 7.1 on Intel Pentium M processor (1.6GHz). The execution time of system sections has shown in the table 2. From

Table 2. Computation time cost

Procedure	Preprocessing	Feature Extraction	Matching
Computation Cost (ms)	1.91	0.03	58.43

the table, the procedure of preprocessing mainly refers to the window ROI extraction described in the subsection 3.1, feature extraction includes the phase of 2-D wavelets decomposition feature extraction and construction of quaternion, and matching time means matching between template and tester with QEPD. It is true that computation time cost is probably reduced provided that we optimize the code.

4.5   A Comparison with Palmprint QEPD

In the experimental section, we also propose palmprint QEPD for a comparison of performance with such finger texture QEPD. There is a distribution of QEPD matching score between genuine and imposter in fig.8.

From the graph, the distance between the peak of inner-class and outter-class is shorter than that of finger texture of QEPD. If two peaks are "closer", it will be more difficult to discriminant genuine and impostor. Here the peak of intra-class is at 0.1, and inter-class 0.24. They are unlikely to separate widely like

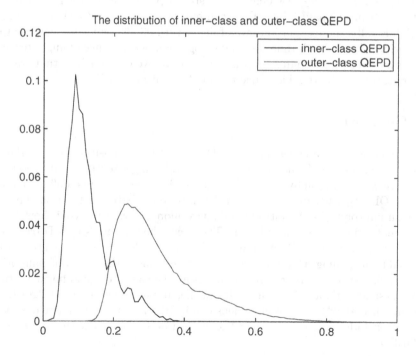

Fig. 8. The distribution of genuine and impostor palmprint QEPD matching score

finger texture QEPD, so as to affect recognition performance. Under the same classifier condition, we propose a comprehensive comparison for two QEPDs as follows.

Table 3. A comparison of palmprint QEPD and finger texture QEPD

	Palmprint QEPD	Finger texture QEPD
Original image size	128×128	12×72
Wavelet decomposition level	3	2
Quaternion matrix size	16×16	3×18
Recognition rate	92.87%	93.36%

This table shows a difference between palmprint QEPD and middle finger texture QEPD. As to original sample size, palmprints in our database are acquired as low resolution 128×128 images, while finger texture images are 12×72, which is less than palmprint's. This means system needs more time to process original palmprint images for wavelet decomposition. On the item of wavelet decomposition level, we here make a choice of level 3 and 2 respectively. Harr kernel is chosen in the decomposition. Here decomposition at Level 2 is faster than level 3. That is to say, this can save time for finger texture, rather than palmprint. Notice that quaternion matrix size is referred as to the size of matrix that stores $\overline{P}P$ or $|\overline{P}Q|$. Evidently, this size of palmprint is larger than finger texture's. It means palmprint QEPD can hardly save time cost as finger texture QEPD. Finally, the recognition rate of finger with 93.36% is higher than that of palmprint. That is, both time cost and recognition performance of finger texture palmprint outperform than palmprint. From data, we can conclude that trait of finger texture is a better biometric than palmprint in QEPD.

5   Conclusion

This paper proposes a novel approach 2D Wavelet Decomposition Feature Parallel Fusion with a QEPD matching score for middle finger texture verification. As 4-feature parallel fusion by wavelets decomposition, the paper first defines a novel finger ROI, finger texture, for wavelet decomposition feature extraction. Then it gives an interpretation by introducing quaternion, in order to avoid shortage of mathematic significance since fusion. Then the scheme defines QEPD based on a term of quaternion QEP, as matching score. Finally, the experimental result is capable of gaining a better recognition rate 93.36% than that of palmprint but at a fast speed. Through this comparison, the size of original finger texture image is smaller than palmprint's. Meanwhile, an intermediate in the algorithm, quaternion matrix has a lower dimensionality than palmprint. Therefore, it is safe to conclude that trait of finger texture is a better biometric than palmprint in QEPD.

Acknowledgment

This paper is fully supported by National Natural Science Funding on No. 60773015 and Research Fund for the Doctoral Program of Higher Education on No. 20060004007.

References

1. Liu, T., Xia, Z., Xie, H.: Information fusion technique and its applications. National Defended Industry Press (1998)
2. Li, Q., Qiu, Z.D., Sun, D.M., Zhang, Y.Q.: Subspace Framework for Feature-Level Fusion with Its Application to Handmetric Verification. In: The 8th International Conference on Signal Processing, vol. 4, pp. 16–20 (2006)
3. Jain, A., Nandakumar, K., Ross, A.: Score normalization in multimodal biometric systems. Pattern Recognition 38(12), 2270–2285 (2005)
4. Sanderson, C., Paliwal, K.K.: Identity verification using speech and face information. Digital Signal Processing 14(5), 449–480 (2004)
5. Lang, F., Zhou, J., Bin, Y., Song, E., Zhong, F.: Quaternion Based Information Parallel Fusion and Its Application in Color Face Detection. In: ICSP 2006 (2006)
6. Ross, A., Jain, A.: Information fusion in biometrics. Pattern Recognition Letters 24(13), 2115–2125 (2003)
7. Yang, J., Yang, J.Y.: A Novel Feature Extraction Method Based on Feature Integration. Chinese Journal of Computer 2002(6), 570–575 (2002)
8. Yang, J., Yang, J.Y., Zhang, D., Lu, J.F.: Feature fusion: parallel strategy vs. serial strateg. Pattern Recognition 36(6), 1369–1381 (2003)
9. Wen, L.L.: Quaternion Matrix, pp. 96–100. National University of Defense Technology Press, ChangSha (2002)

Studies of Fingerprint Template Selection and Update

Yong Li, Jianping Yin, En Zhu, Chunfeng Hu, and Hui Chen

School of Computer, National University of Defense Technology, Changsha, 410073, China
ylee.nudt@gmail.com, JPYin@nudt.edu.cn, nudt_EN@263.net,
huchfeng@gmail.com, chenhui_sunrise@tom.com

Abstract. A fingerprint recognition procedure usually contains two stages: registration and authentication. Most fingerprint recognition systems capture multiple samples of the same finger (e.g., eight impressions of a person's left index finger) at the stage of registration. As a result, it is essential to select several samples as templates. This paper proposes two algorithms maximum match scores (MMS) and greedy maximum match scores (GMMS) based on match scores for template selection. The proposed algorithms need not involve the specific details about the biometric data. Therefore, they are more flexible and can be used in various biometric systems. The two algorithms are compared with Random and sMDIST on the database of FVC2006DB1A, and the experimental results show that the proposed approaches can improve the accuracy of biometric system efficiently. Based on the maximized score model, we propose two strategies: ONLINE and OFFLINE for templates update and analyze the relationship between the two strategies. Preliminary experiments demonstrate that OFFLINE strategy gains better performance and GMMS performs better and can gain steady improvement.

Keywords: biometric template, template selection, template update, match score, fingerprint recognition.

1 Introduction

Biometric recognition refers to the use of distinctive physiological or behavioral characteristics for automatically confirming the identity of a person. It is becoming apparent that unibiometric system is not sufficient to meet the requirement of system which need very high performance and security. Multibiometrics which combines more information is expected to alleviate the limitations of unibiometric systems [1]. Fingerprint recognition is the most popular and reliable biometric technique for automatic personal identification and plays an important role in multibiometric system from application to research activities. Biometric systems usually include two stages: registration and authentication. In practice, most biometric systems will capture multiple samples of the same biometric trait at the stage of registration. For example, some fingerprint systems require the user to enroll eight impressions into the system. There are three techniques to deal with the multiple samples literally. The first method [3,4,11] is template selection which chooses several samples as templates

T.-h. Kim et al. (Eds.): FGCN 2008 Workshops and Symposia, CCIS 28, pp. 150–163, 2009.

from the multiple samples and extracts feature from each template during Enrollment. At the authentication stage, the inquiry fingerprint is matched with these templates and then score level fusion or decision level fusion is used to give the final decision. The second method [6,8] is to combine all the samples as a super template and extract features from it. The third method [2,5,7] is to extract features from each sample first and then combine the features together as one feature template. With the development of computer hardware and software especially for storage and parallel processing technique, storing multiple templates is feasible. Techniques on combining image files or features mostly used in fingerprint recognition may not suitable for other biometric recognition systems. Also, such techniques need deep comprehension of feature extraction and match details. In addition, it is difficult to combine too many raw biometric samples or feature templates together. So it is helpful to select template based on match scores after enrollment.

Template selection will improve the overall system's accuracy efficiently if enough samples are captured in enrollment. However, it is not user-friendly to capture a number of fingerprints of the same finger at long intervals in the registration phase. During the authentication stage of a fingerprint recognition system, input fingerprints are successively received and compared with the templates. If an input fingerprint is successfully matched with a template, these two fingerprints are verified to originate from the same finger. Therefore, input fingerprints can be used to update the matched template. And score based template update is very important to improve the match performance of biometric systems [3,12,13].

The rest of the paper is organized as follows. In Section 2 the model of templates selection has been studied and the relation between sample number and template number has been analyzed especially for the number of samples equal to 8. Based on match scores, two algorithms have been described for template selection in section 3. In section 4, two strategies have been proposed for template update. To study the effectiveness of the proposed technique, section 5 gives the experimental results. The last section summarizes the results of this work and provides future directions for research.

2 Multiple Templates Fusion

The match score is a measure of similarity between the template and the input biometric feature vectors. Combining different match scores with fusion strategy to give the final decision is called score level fusion.

In daily life, people recognize each other usually by face and voice. In different environments and conditions, we may often find that the similarity between two persons is different. Although the probability that a large number of minutiae from impressions of two different fingers will match is extremely small, fingerprint matchers aim to find the "best"alignment [14]. A fingerprint system compares the template and input biometric traits in different angles and gives a match score for each angle. Finally, the greatest mark is chosen as the result which represents the greatest similarity of the two fingerprints. We call this principle the greatest similarity principle.

The problem of multi-template system can be described as follows:

Given N fingerprint templates T_i (i=1,2,...N) which are N templates from N the same finger, and the input fingerprint I_1. N match scores s_i(i=1,2,...N) represent the similarity between T_i(i=1,2,...N) and I_1. We assume that the N templates from person A and the input fingerprints from person B. We need to identify if B and A are the same person. In our experiments, all the match scores from the same matcher are between 0 and 1, so we need not transform the match scores. Different from Kittler' rule which is based on Bayesian decision theory, we try to find the fusion rules based on similarity between A and B based on the N score marks. That is to say we try to find a new match score which represents similarity between A and B based on the N match scores. According to the greatest similarity model, the match score between A and B is apt to be bigger.

There are usually three kinds of score fusion rules: density based rules, transformation based rules and classifier based rules [10]. In this paper, we pay attention to transformation based rules which are also called fixed fusion rules. Kittler et al. [15] proposed a set of fusion strategies namely, sum, product, min, max, median and majority vote rules. We study the fusion rules based on the similarity model. In our paper, the definition of the five fixed fusion rules: sum, max, min, median and product are as follows.

sum:

$$S_{sum} = \frac{1}{N}\sum_{i=1}^{N} s_i \tag{1}$$

product:

$$S_{product} = (\prod_{i=1}^{N} s_i)^{1/N} \tag{2}$$

max:

$$S_{max} = \max_{i=1}^{N}(s_i) \tag{3}$$

min:

$$S_{min} = \min_{i=1}^{N}(s_i) \tag{4}$$

median:

$$S_{median} = \underset{i=1}{\overset{N}{median}}(s_i) \tag{5}$$

It is not difficult to prove (6).

$$S_{max} \geq S_{sum} \geq S_{product} \geq S_{min} \text{ and } S_{max} \geq S_{median} \geq S_{min} \tag{6}$$

If $s_i \in [0,1], i = 1, 2, \ldots N$, then the final marks after fusion are between 0 and 1. For a two-class classification problem, score based max rule is totally the same as decision level rule OR. Also, min rule is the same as AND, and median rule is the same as majority vote rule. According to the greatest similarity principle and expression (6), max rule may perform better than other rules. In our experiment we find that max rule has a better performance than other rules.

3 Multiple Templates Selection Analyze

3.1 Templates Selection and Multibiometrics

Biometric recognition systems usually keep several templates and execute multiple matches in authentication for an individual. The system validates a person's identity by comparing the captured biometric data with his own biometric templates stored in the database. Systems which keep more than one template for each individual may be classified as a kind of multisample system. Based on the theory of multibiometrics, multiple templates can gain high matching accuracy. Consider a two-class classification problem and a multiclassifier system consisting of N classifiers (assume N is odd); the majority vote rule classifies an input pattern as belonging to the class that obtains at least K=(N+1)/2 votes. If p is the probability that a single classifier performs correctly, then the probability of the multiclassifier is as follows [9]:

$$P(N) = \sum_{m=K}^{N} \binom{N}{m} p^m (1-p)^{N-m} \tag{7}$$

The formula (1) assumes that the classifiers themselves are statistically independent. However, multiple templates come from the same biometric trait, which can not be independent in practice. Anyway, multiple templates can improve the accuracy of biometric system, but how many templates are suitable and how do multiple templates achieve better performance can be a problem. In this paper, we studied the two problems and got some useful conclusions.

3.2 A Model of Template Selection

The problem of template selection may be posed as follows: Choose K(K<N) biometric templates from N biometric samples captured from users during the enrollment which can represent the variability as well as the typicality observed in the N biometric samples best.

Definition 1: During the enrollment, the user inputs N biometric samples $i_1, i_2, \ldots, i_N$. After feature extraction, N candidate feature templates $t_1, t_2, \ldots, t_N$ are available. Score matrix $S_{N \times N}$ is constructed in which the item $s_{m,n}$ represents the match score between the template t_m and the input sample i_n. In our experiment, all the match scores are between 0 and 1. It should be noticed that $s_{m,n}$ may not be equal to $s_{n,m}$ because the feature template has less information than its sample after feature extraction.

Definition 2: The distance between biometric feature templates set A and biometric samples set B is:

$$d(A,B) = \sum_{\forall t_m \in A, \forall i_n \in B} (1 - s_{m,n})$$

(8)

The K templates selected is denoted as template set T_K and the left N-K samples is denoted as input set I_{N-K}. For the future inputs $i_{N+1}, i_{N+2}, \ldots$ denoted as set I_N^+, our aim is that the distance between T_K and I_N^+ denoted as $d(T_k, I_{N-K})$ should be minimized. However, the future inputs are not available. So replace the set I_N^+ with the left samples set I_{N-K}, the problem is now to choose the K templates that make $d(T_k, I_{N-K})$ minimized. We define the match score between set A and set B as follows:

$$s(A,B) = \sum_{\forall t_m \in A, \forall i_n \in B} s_{m,n}$$

(9)

Finally, substituting formula (9) into formula (8), we obtain formula (10) where $N(A)$ and $N(B)$ refers to the numbers of elements in set A and B.

$$d(A,B) = N(A) * N(B) - s(A,B)$$

(10)

Finally, we obtain the follow formula:

$$d(T_k, I_{N-K}) = K * (N - K) - s(T_k, I_{N-K})$$

(11)

K(N-K) is constant in template selection. Therefore the model is to select K templates from N input samples which make $s(T_k, I_{N-K})$ maximized. To express simply, we call the model maximized score model and record the maximized $s(T_K, I_{N-K})$ as $Max(T_K)$.

3.3 How Many Templates Should Be Selected?

It is a key issue that how many templates should be selected. There are at least three factors to be considered to decide the value of K. First, the aim of template selection is to reduce the number of templates and to keep good performance at the same time. Second, although we say that storage and parallel processing technique progress greatly, the memory space and CPU speed is not unlimited and more space and higher CPU speed means more cost. We should make a tradeoff between the templates number and the cost. Third, according to the maximized score model, we should make the $s(T_K, I_{N-K})$ maximized on condition that I_{N-K} represents the future inputs set I_N^+. Therefore, N-K should be large enough so that I_{N-K} can represent I_N^+. In practice, $s_{m,n}$ and $s_{n,m}$ are usually very similar. So given $s_{m,n} = s_{n,m}$, $Max(T_K) = Max(T_{N-K})$. That means choose K templates are the same with N-K templates according to the maximized score model. So we suggest $K \leq [N/2]$ if N is large enough. Also, from Table 1 in section 5, we can find that the EER is decreased begin slowly when $K \geq 4$. In our experiment, 3~5 templates are suitable when N=8.

4 Score Based Template Selection

The common template selection strategy is the random selection strategy (Random). This strategy just randomly chooses several templates each time. Though it may be poor performance, it is useful to be compared with other strategies. According to maximized score model mentioned in section 2, we want to choose the K templates that make $s(T_K, I_{N-K})$ maximized. Based on this model, we propose two algorithms. In the following algorithms Table 1, $S_{N \times N}$ delegates the match score matrix. The value of $s_{m,n}$ is the match score of template t_m matched with input sample i_n. Choose[K] is used to record the templates selected.

Table 1. Maximum Match Scores (MMS)

1	Initialize N, K, $S_{N \times N}$, Choose[K],
2	list all T_K^i and T_{N-K}^i $(1 \le i \le C_N^K)$
3	for $i \leftarrow 1$ to C_N^K do begin
4	begin
	using
5	$S_{N \times N}$ compute $s_i \leftarrow s(T_K^i, T_{N-K}^i)$ (1)
6	end
7	find i^* $(i^* \in 1, \ldots, C_N^K)$
8	such that $s_{i^*} \ge s_i$ $(i = 1, \ldots, C_N^K)$
9	record $T_K^{i^*}$ with Choose[K]

To analyze the line (5) of MMS algorithm, we can get the complexity of MMS as (12):

$$(1/4)N^2 C_N^{N/2} \ge (1/4)N^2 \sqrt{2}^N \qquad (12)$$

When N increases, the complexity will increase exponentially. Then we propose another algorithm Greedy Maximum Match Scores (GMMS) as Table 2.

The idea of algorithm GMMS just can be described as choosing one sample which has maximum match scores with the left samples each time. The complexity of GMMS is $O(N^2)$. Umut Uludag [3] proposed two methods DEND and MDIST to select templates. From his work, the MDIST method has a better performance. The distance in his work is obtained by matching the minutiae point sets of the two fingerprint impressions, and we call MDIST based on match scores sMDIST. The basic idea of MMS, GMMS and sMDIST is to choose the partition of T_N that makes the sum of match scores maximized.

Table 2. Greedy Maximum Match Scores (GMMS)

1   Initialize N, K, $S_{N \times N}$, Choose[K],
2   for $i \leftarrow 1$ to K do
3   begin
4   comput sum(j)= $\displaystyle\sum_{m=1,m \neq j}^{N} s_{j,m}$ $j = 1, \ldots N$
5   find $j*$ ($j* \in 1, \ldots N$)
6   such that sum($j*$) $\geq sum(j)$ $j = 1, \ldots N$
7    $Choose[i] \leftarrow j*$
8    $s_{j*,m} \leftarrow 0, s_{m,j*} \leftarrow 0, m = 1, \ldots N$
9   end

5 Score Based Template Update

5.1 Introduction to Template Update

During enrollment, N samples are captured at the same time. However, to make system user-friendly and save time, N is small usually. Also, it may be not a good strategy to capture too many samples during one time because the samples will not vary of wide scope. Biometric trait varies along with age and suddenness. For example, older people will have more wrinkles than young people and fingerprints from older people may have more cut than young people. Therefore, template update is very important in biometric recognition system [3,12,13]. Umut Uludag [3] proposed two strategies BATCH-UPDATE and AUGMENT-UPDATE for template update. And in his paper [3], the experimental results show that   AUGMENT-UPDATE gains better performance than BATCH-UPDATE. This is because biometric samples are varying gradually not abruptly. Also feature level fusion [2,5,7] and image level fusion [6,8] are used in template update and these two strategies are not very different with template selection. In our paper, template update means to update the K templates with the successively accept samples a_i during authentication. In order to conduct template update in authentication, biometric system need to store the match scores among the K templates. As we assume that the biometric database just store the feature exacted from raw data, we cannot get the match score between template raw data i_k and the successively accept sample a_i during authentication. Therefore, we don't distinguish between $s_{m,n}$ and $s_{n,m}$ in template update.

5.2 Two Strategies of Template Update

The strategies of template update can be divided into two classes. One is ONLINE update which activates the update procedure whenever a new biometric sample is accepted. The other is OFFLINE update which activates the update procedure during a fixed period. And we just call the two methods: ONLINE and OFFLINE. On the one hand, ONLINE strategy need no more space for update procedure and can update as soon as a new sample accepted, but it may time consuming and will not gain the global benefits. On the other hand, OFFLINE strategy needs space to store the feature template until update procedure is invoked and cannot update as soon as possible, but it may update globally and may cost less time. For OFFLINE strategy, we use GMMS, MMS and sMDIST methods to verify the improvement of performance. For ONLINE strategy, we also use the three methods after each authentication.

6   Experimental Results

The experiment is conducted on the database FVC2006 DB1A [16]which has 140 fingers and each finger has 12 impressions. The front 8 impressions of each finger are used for template selection and the last 4 impressions used as the future inputs for authentication.

6.1   Experiment for Template Selection

When we conduct experiment for template selection, we first select K templates based on match scores using different strategies and then execute K matches for each authentication. At last, the K match scores are fused with max rule [10] to do final decision. In our experiment we find that max rule has a better performance than other rules.

When K=1, MMS, GMMS and sMDIST are the same. Figure 1 shows the comparison between MMS and Random when K=1. From figure 1, we can find that template selection can improve the performance efficiently.

Figure 2 shows EER of the four algorithms with different K(N=8). From EER aspects, MMS gains better performance than GMMS and other two methods. When K is closer to N, the performance of all the methods will be closer to the performance with Random selection strategy.

As shown from Figure 3 to Figure 10, all the three algorithms MMS, GMMS and sMDIST gain better performance than Random. From Figure 3 and Figure 4, we can see that when K=2, sMDIST is a bit better than MMS and GMMS. Then from Figure 5 to Figure 10, we can find that MMS and GMMS are better than sMDIST when K>2.

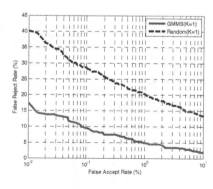
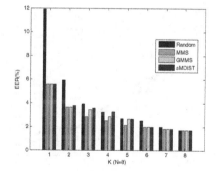

Fig. 1. ROC curves show comparison between MMS and Random when K=1(N=8)

Fig. 2. EER(%) of the four algorithms: Random, MMS, GMMS, sMDIST with different K(N=8)

158   Y. Li et al.

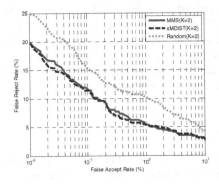

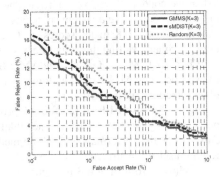

Fig. 3. ROC curves show comparison between MMS , sMDIST and Random when K=2

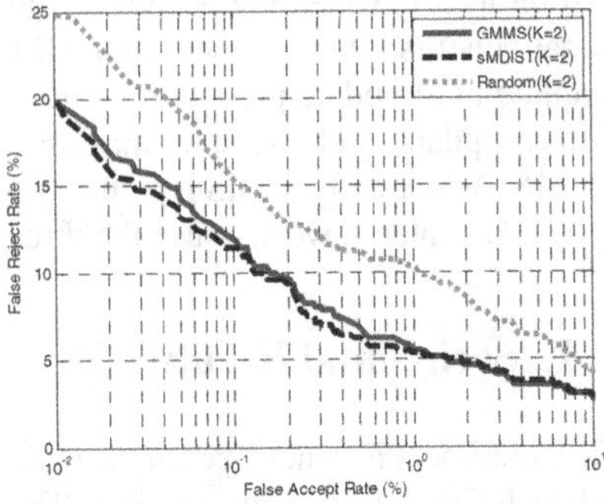

Fig. 4. ROC curves show comparison between GMMS, sMDIST and Random when K=2

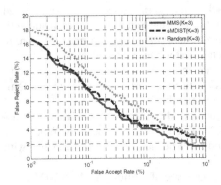

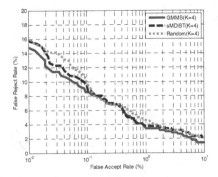

Fig. 5. ROC curves show comparison between MMS , sMDIST and Random when K=3

Fig. 6. ROC curves show comparison between GMMS, sMDIST and Random when K=3

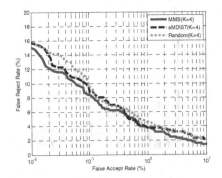

Fig. 7. ROC curves show comparison between MMS , sMDIST and Random when K=4

Fig. 8. ROC curves show comparison between GMMS, sMDIST and Random when K=4

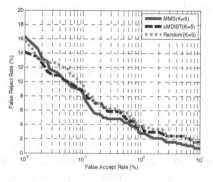

Fig. 9. ROC curves show comparison between MMS , sMDIST and Random when K=5

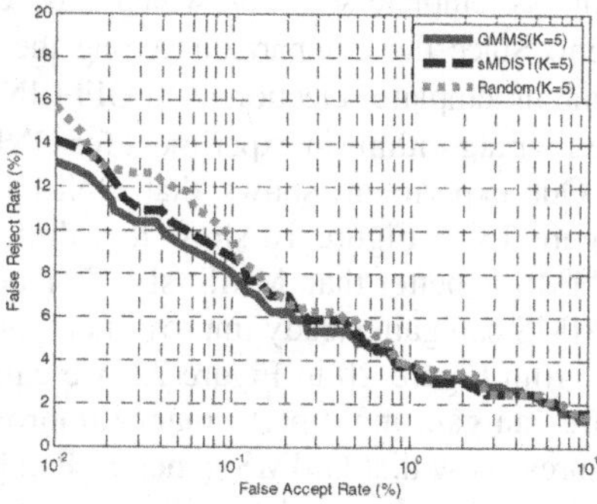

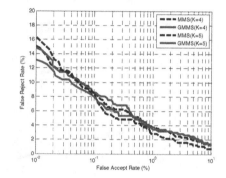

Fig. 10. ROCs show comparison between GMMS, sMDIST and Random when K=5

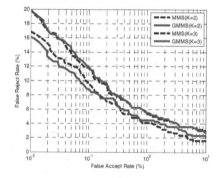

Fig. 11. ROCs show comparison between MMS and GMMS when K=2,3

Fig. 12. ROCs show comparison between MMS and GMMS when K=4,5

From Figure 11 and Figure 12, we can find the facts that when FAR below a special threshold GMMS is better than MMS and when FAR above the threshold MMS is better than GMMS. Although it is the optimized method according to the maximized score model, MMS is not always better than GMMS. The reason may be that N=8 is not large enough and N-K can not represent the set T_N^+ effectively. In addition, consider each biometric trait as one class, the maximized score model makes the intra-class distance smaller and at the same time may make the inter-class distance smaller too.

6.2 Experiment for Template Update

When we conduct experiment for template update, we first choose 4 templates as the template group. Then the next four samples are used for template update. For ONLINE strategy, we conduct template update procedure four times. Each time, we choose four templates from the original four templates and the newly added samples. So the template update procedure is based on the template selection procedure. For OFFLINE strategy, we conduct template update procedure only one time. This procedure is the same as template selection which chooses 4 templates from 8 templates in our experiment. Since the comparison among the MMS, GMMS, Random and sMDIST is the same in template selection as in OFFLINE template update procedure, we just compare the four algorithms in experiment for ONLINE template update.

Our experiment shows that ONLINE update is more complex than OFFLINE. Figure 13 to Figure 16 show that during ONLINE update procedure, the algorithm GMMS is better than MMS, sMDIST and Random. Figure 17 to Figure 19 show that GMMS can gain steady improvement than the other two algorithms.

From Figure 20 to Figure 22, we can find that OFFLINE templates update procedure can gain more performance improvement than ONLINE procedure. Also, the 3 figures show that GMMS is better than MMS. From Figure 21 and Figure 22, we also find that template update will not always improve the performance of biometric systems. It is crucial for designers to ensure the efficiency of algorithms which used in ONLINE and OFFLINE template update procedure.

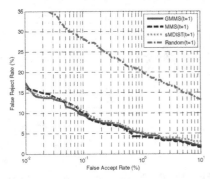

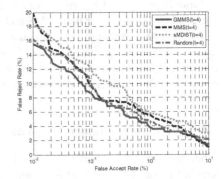

Fig. 13. ROC curves show comparison among MMS, GMMS, sMDIST and Random after t=1 time update

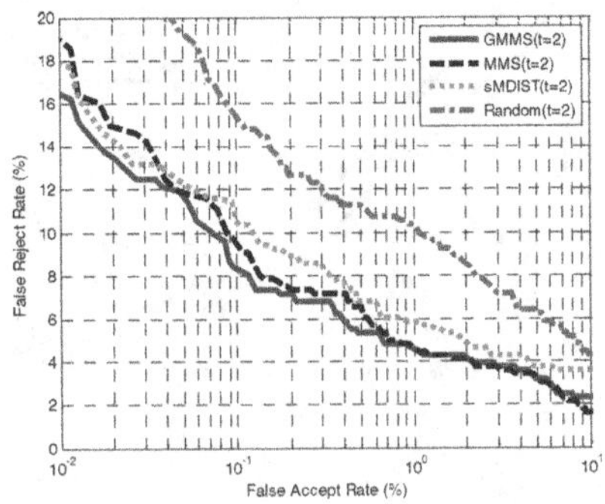

Fig. 14. ROC curves show comparison among MMS, GMMS, sMDIST and Random after t=2 times update

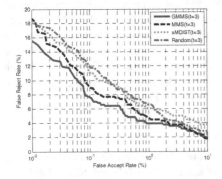

Fig. 15. ROC curves show comparison among MMS, GMMS, sMDIST and Random after t=3 times update

Fig. 16. ROC curves show comparison among MMS, GMMS, sMDIST and Random after t=4 times update

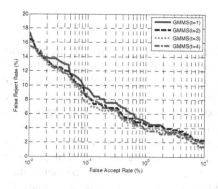

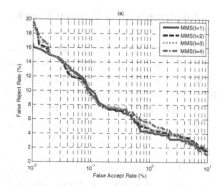

Fig. 17. ROC curves show the performance after t times update using the GMMS in ONLINE update

Fig. 18. ROC curves show the performance after t times update using the MMS in ONLINE update

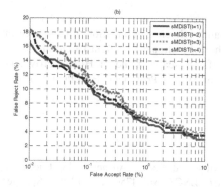

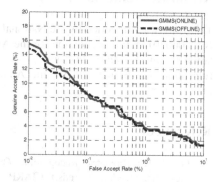

Fig. 19. ROC curves show the performance after t times update using the sMDIST in ONLINE update

Fig. 20. ROC curves show the performance between ONLINE and OFFLINE templates update with GMMS

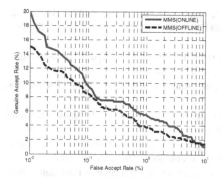

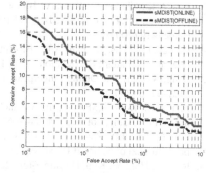

Fig. 21. ROC curves show the performance between ONLINE and OFFLINE templates update with MMS

Fig. 22. ROC curves show the performance between ONLINE and OFFLINE templates update with sMDIST

7 Summary and Future Work

A systematic procedure for template selection and update is critical to the perform-ance of a biometric system. Even in which system combines raw data or features to create a new template may do template selection first and then combines the selected templates. In this paper, we propose two algorithms based on match scores. The ex-perimental results show that the two methods: MMS and GMMS can improve the accuracy of biometric system efficiently. Our techniques do not need to evolve with the details of feature extraction and match, so it is convenient to use the method in other biometric systems. Currently, our fingerprint database just has 12 impressions for each individual. In the future, we prepare to experiment on other biometric data-base which has more samples for each individual. Also, we prepare to study the ONLINE template update procedure further.

Acknowledgements

This work was supported by the National Natural Science Foundation of China (Grant No. 60603015), the Foundation for the Author of National Excellent Doctoral Disser-tation (Grant No. 2007B4), and the Scientific Research Fund of Hunan Provincial Education (the Foundation for the Author of Hunan Provincial Excellent Doctoral Dissertation).

References

1. Jain, A.K.: Biometric Recognition: Overview and Recent Advances. In: Rueda, L., Mery, D., Kittler, J. (eds.) CIARP 2007. LNCS, vol. 4756, pp. 13–19. Springer, Heidelberg (2007)
2. Ryu, C., Hakil, K., Jain, A.K.: Template adaptation based fingerprint verification. In: Proc. of International Conference on Pattern Recognition, ICPR, Hong Kong, August 2006, vol. 4, pp. 582–585 (2006)
3. Jain, A.K., Uludag, U., Ross, A.: Biometric template selection: a case study in fingerprints. In: Kittler, J., Nixon, M.S. (eds.) AVBPA 2003. LNCS, vol. 2688, pp. 335–342. Springer, Heidelberg (2003)
4. Uludag, U., Ross, A., Jain, A.K.: Biometric template selection and update: a case study in fingerprints. Pattern Recognition 37(7), 1533–1542 (2004)
5. Jiang, X., Ser, W.: Online Fingerprint Template Improvement. IEEE Trans. PAMI 24(8), 1121–1126 (2002)
6. Jain, A.K., Ross, A.: Fingerprint Mosaicking. In: Proc. Int'l Conf. on Acoustic Speech and Signal Processing, vol. 4, pp. 4064–4067 (2002)
7. Zhu, E., Yin, J.P., Zhang, G.M., Hu, C.F.: Merging Features of Multiple Template Finger-print. Journal of National University of Defense Technology 27(6), 26–29 (2005)
8. Ryu, C., Han, Y., Kim, H.: Super-template Generation Using Successive Bayesian Estima-tion for Fingerprint Enrollment. In: Kanade, T., Jain, A., Ratha, N.K. (eds.) AVBPA 2005. LNCS, vol. 3546, pp. 710–719. Springer, Heidelberg (2005)
9. Maltoni, D., Maio, D., Jain, A.K., Prabhakar, S.: Handbook of Fingerprint Recognition, pp. 235–237. Springer, Heidelberg (2003)

10. Ross, A., Nandakumar, D., Jain, A.K.: Handbook of Multibiometrics. Springer, Heidelberg (2006)
11. Lumini, A., Nanni, L.: A clustering method for automatic biometric template selection. Pattern Recognition 39, 495–497 (2006)
12. Roli, F., Didaci, L., Marcialis, G.L.: Template Co-update in Multimodal Biometric Systems. In: Lee, S.-W., Li, S.Z. (eds.) ICB 2007. LNCS, vol. 4642, pp. 1194–1202. Springer, Heidelberg (2007)
13. Aggarwal, G., Ratha, N.K., Bolle, R.M., Chellappa, R.: Multi-biometric cohort analysis for biometric fusion. In: Acoustics, Speech and Signal Processing, ICASSP 2008, pp. 5224–5227 (2008)
14. Maltoni, D., Maio, D., Jain, A.K., Prabhakar, S.: Handbook of Fingerprint Recognition. Springer, Heidelberg (2006)
15. Kittler, J., Hatef, M., Duin, R.P.W., Matas, J.: On combining classifiers. IEEE Trans. on Pattern Anal. Machine Intell. 20(3), 226–239 (1998)
16. http://bias.csr.unibo.it/fvc2006/

Characterizing Genes by Marginal Expression Distribution

Edward Wijaya, Hajime Harada, and Paul Horton

Computational Biology Research Center
AIST Waterfront, Bio-IT Research Building
2-42 Aomi, Koto-ku, Tokyo 135-0064
horton-p@aist.go.jp

Abstract. We report the results of fitting mixture models to the distribution of expression values for individual genes over a broad range of normal tissues, which we call the *marginal distribution of the gene*. The base distributions used were normal, lognormal and gamma. The expectation-maximization algorithm was used to learn the model parameters. Experiments with artificial data were performed to ascertain the robustness of learning. Applying the procedure to data from two publicly available microarray datasets, we conclude that lognormal performed the best function for modeling the marginal distributions of gene expression. Our results should provide guidances in the development of informed priors or gene specific normalization for use with gene network inference algorithms.

Keywords: microarray, marginal distributions, mixture models.

1 Introduction

Several studies have used finite mixture to model the distributions of gene expression values. Some notable works include those by Hoyle [1] and Yuan [2]. Hoyle investigated the entire distributions of expression levels of mRNA extracted from human tissues. Yuan examined the distribution of gene expression's correlation coefficient on cancer cells. However, less analysis has been done on the marginal distribution of gene expression levels.

In this paper we present a preliminary analysis of modeling the marginal distributions using mixture models with normal, lognormal or gamma distributions as the model components. Compared the previous works this study attempt to answer the following questions:

1. Is there a generic form of distribution that describe best the marginal expression of genes?
2. Can we find what is common amongst the genes that have similar mixture components?

The gamma and lognormal distributions belong to family of skewed distributions. We expected that these distributions could model the microarray data

T.-h. Kim et al. (Eds.): FGCN 2008 Workshops and Symposia, CCIS 28, pp. 164–175, 2009.

that is often skewed [3]. Additionally we use the standard normal distribution as a control experiments on which we compare how well gamma and lognormal mixture models perform.

Our choice of using gamma distribution is because of its flexible shape. Furthermore it has been successfully used in many studies of biological systems [4,5,6]. With regard to the lognormal distribution, there is a strong evidence that this distribution appears in many biological phenomena [7]. In practice it is also convenient for analyzing microarray data is because it is easy to perform calculations and capable of determining the data z-scores, a possible common unit for data comparison [8]. Below we describes the detail of our methods and experimental results.

2   Methods

2.1   Statistical Model

Let $\{x_i\}, i = 1, \ldots, N$ denote the expression value of a gene probe, where N is the total number of observations (samples). Under a mixture model, the probability density function for observing finite data points x_i is:

$$p(x) = \sum_{j=1}^{K} p(x|j)P(j) \tag{1}$$

The density function for each component is denoted as $p(x|j)$. In appendix we give the formal description of density function from three types of distributions used in our model. And $P(j)$ denotes the prior probability of the data point having been generated from component j of the mixture. These priors are chosen to satisfy the constraints $\sum_{j=1}^{K} P(j) = 1$. The log likelihood function of the data is given by:

$$LL = -logL = -\sum_{i=1}^{N} log \sum_{j=1}^{K} p(x_i|j)P(j) \tag{2}$$

We use expectation-maximization (EM) algorithm [9] to learn mixture models of normal, lognormal and gamma distribution for each probe's expression level. It is implemented with **R** programming language. The EM algorithm iteratively maximizes the loglikelihood and update the conditional probability that x comes from K-th component. This is defined as

$$p(x|j)^* = E[p(x|j)|x, \hat{\theta}_{A_1}, \hat{\theta}_{B_1}, \ldots, \hat{\theta}_{A_K}, \hat{\theta}_{B_K}] \tag{3}$$

The set of parameter $[\hat{\theta}_{A_1}, \hat{\theta}_{B_1}, \ldots, \hat{\theta}_{A_K}, \hat{\theta}_{B_K}]$ is a maximizer of loglikelihood, for given $p(x|j)$. The EM algorithm iterates between an E-step where values $p(x|j)^*$ are computed from the current parameter estimates, and M-step in which the loglikelihood with each $p(x|j)$ replaced by its current conditional expectation $p(x|j)^*$ is maximized with respect to the parameters θ_A and θ_B. These two parameters correspond to the parameters in each distributions (i.e. *shape, scale* in gamma and *mean, standard deviation* in normal and lognormal distributions). The detail of the EM algorithm is as follows:

1. Initialize θ_A and θ_B. It is done by first randomly partitioning the dataset into K groups and then calculate *method of moment estimates* for each of the groups.
2. M-step: Given $p(x|j)^*$, maximize loglikelihood (LL) with respect to the parameters θ_A and θ_B. We obtain maximizer for $[\hat{\theta}_{A_1}, \hat{\theta}_{B_1}, \ldots, \hat{\theta}_{A_K}, \hat{\theta}_{B_K}]$ numerically.
3. E-step: Given the parameter estimates from M-step, we compute:

$$p(x|j)^* = E[p(x|j)|x, \hat{\theta}_{A_1}, \hat{\theta}_{B_1}, \ldots, \hat{\theta}_{A_K}, \hat{\theta}_{B_K}] \tag{4}$$

$$= \frac{\hat{p}(x|j)}{\sum_{j=1}^{K} \hat{p}(x|j)} \tag{5}$$

4. Repeat M-step and E-step until the change in the value of the loglikelihood (LL) is negligible.

In order to avoid local maxima, we run the above EM algorithm ten times with different starting points.

2.2   Model Selection

When fitting mixture models to expression data, it is necessary to desierable to choose an appropriate number of components, which fits the data well but does not overfit. For this task we tried two information criteria: AIC (Akaike Information Criterion [10]) and BIC (Bayesian Information Criterion [11]). Specifically:

$$AIC = -2LL + 2c \tag{6}$$

and

$$BIC = -2LL + c \, log(N) \tag{7}$$

To choose models, we fit mixture models with the EM algorithm for one to five components and chose the model with the smallest information criteria value (the degree of freedom c in the above formulas, is equal to $3K - 1$ for K components).

3   Experimental Results

3.1   Estimating Number of Components

We generated simulated datasets from mixture models containing with one, two and three components. We performed two sets of equivalent experiments, one using the gamma and one using the lognormal distribution for the mixture model components. For the component parameters, each distinct combination of the

Table 1. The parameter pairs used as components for the simulated data generating mixture models are shown

Parameters	1	2	3	4
θ_A	0.1	0.1	5	10
θ_B	7	14	3	6

parameter pairs shown in Table 1 was tried; yielding 4 mixture models with one and three components and $6 = \binom{4}{2}$ mixture models with two components. The mixing components ($P(j)$ in equation 1) were set to the uniform distribution in each case. To aid comparison with the real data described below, we simulated datasets with sizes 122 and 158.

Figure 1 and 2. indicates the observed effectiveness of information criteria to induce the number of components in the generating model, from the generated data. BIC outperforms AIC, correctly inducing the number of components in about 80% of the trials. Even for BIC, the error is slightly skewed toward overpredicting the number of components. This suggests it may be possible to further optimize the criteria for this task, but we did not pursue this possibility.

3.2 Real Dataset

As a preliminary study, we investigated the gene expression dataset from human (GDS596) and mouse (GDS592) from GEO database [12] gene expression data repository (http://www.ncbi.nlm.nih.gov/geo/). GDS596 contains data from a study profiling 158 types of normal human tissue (22,283 probes) and GDS592 with 122 types of mouse tissues (31,373 probes) [13].

Likelihood and K-S Test Comparison. We evaluate the likelihood of mixture models from three types of distributions on the real datasets. Furthermore we evaluate goodness of fit of a model by using Kolmogorov-Smirnov (K-S) test as represented by D, the maximum discrepancy in the cumulative probability distribution, and a p-value statistic.

The goodness of fit p-value statistics, indicate that the gamma mixtures can fit the marginal distribution of gene expression reasonable well. However, lognormal mixtures fit better than gamma mixtures. Over all experiments they obtain a higher likelihood than the gamma mixtures. The K-S test also confirm this

Table 2. loglikelihood, D and p-value, averaged over each probe of the GDS596 dataset

#Comp	Normal			Lognormal			Gamma		
	Loglik	D	p-value	Loglik	D	p-value	Loglik	D	p-value
1	-1063.02	1.04e-3	0.99	-212.85	2.30e-4	0.99	-968.59	3.00e-3	0.99
2	-979.13	1.74e-3	0.99	-205.58	4.55e-4	0.99	-955.11	1.60e-3	0.99
3	-952.29	2.72e-3	0.99	-204.55	7.04e-4	0.99	-963.20	1.19e-3	0.99
4	-913.66	4.63e-2	0.92	-203.05	8.27e-4	0.99	-967.20	9.88e-4	0.99
5	-881.67	1.55e-2	0.90	-201.78	1.26e-3	0.99	-968.56	5.77e-4	0.99

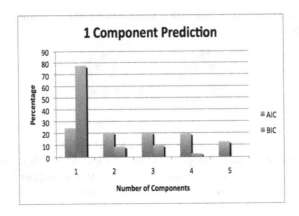

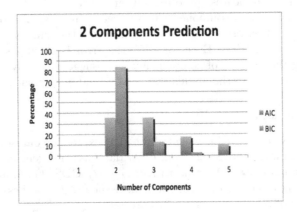

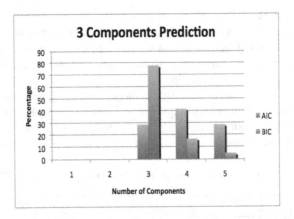

Fig. 1. Performance of lognormal mixture models in predicting number of components on simulated datasets. The percentage of accuracy is taken from the average performance of 4 or 6 parameter settings x 100 trials x 2 dataset sizes.

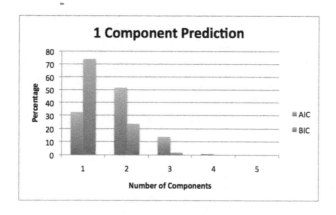

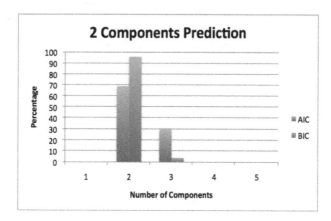

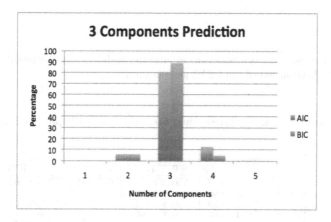

Fig. 2. Performance of gamma mixture models in predicting number of components on simulated datasets. The percentage of accuracy is taken from the average performance of 4 or 6 parameter settings x 100 trials x 2 dataset sizes.

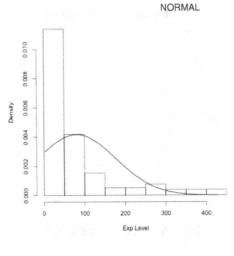

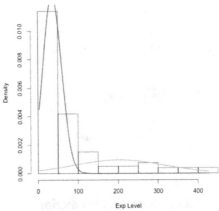

Fig. 3. The components (one and two) of mixture models of normal distribution fit to the marginal distribution of the CSN3 gene is shown

Table 3. Loglikelihood, D and p-value, averaged over each probe of the GDS592 dataset

#Comp	Normal			Lognormal			Gamma		
	Loglik	D	p-value	Loglik	D	p-value	Loglik	D	p-value
1	-728.47	1.14e-3	0.99	-85.04	7.70e-5	0.99	-492.66	1.72e-3	0.99
2	-683.53	1.44e-3	0.99	-47.39	7.73e-5	0.99	-437.79	1.05e-3	0.99
3	-661.46	2.70e-3	0.99	-46.64	9.20e-5	0.99	-487.58	1.26e-3	0.99
4	-637.21	5.79e-3	0.97	-46.01	6.17e-5	0.99	-474.41	1.19e-3	0.99
5	-612.37	9.44e-2	0.88	-45.08	7.45e-5	0.99	-465.15	8.36e-4	0.99

LOGNORMAL

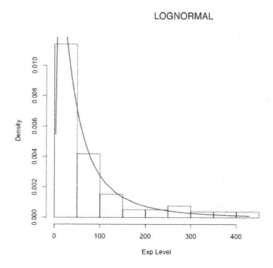

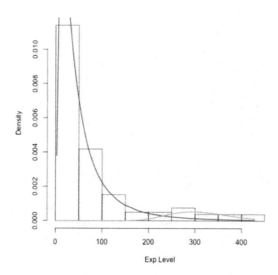

Fig. 4. The components (one and two) of mixture models of lognormal distribution fit to the marginal distribution of the CSN3 gene is shown

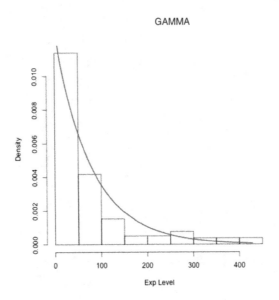

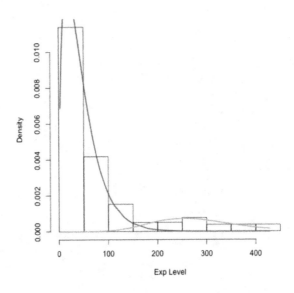

Fig. 5. The components (one and two) of mixture models of gamma distribution fit to the marginal distribution of the CSN3 gene is shown

Table 4. Statistically overrepresented terms from the two-component probes in GDS596

Terms	GO	p-value
cytoskeleton	GO:0005856	4.28e-3
non-membrane-bound organelle	GO:0043228	8.3e-3
organelle part	GO:0044422	6.33e-2
intracellular	GO:0005622	**1.17e-1**
cytoplasm	GO:0005737	1.75e-1

Table 5. Statistically overrepresented terms from the two-component probes in GDS592

Terms	GO	p-value
proteinaceous extracellular matrix	GO:0005578	6.18e-2
intracellular	GO:0005622	**5.48e-2**
midbody	GO:0030496	9.85e-2
cell soma	GO:0043025	9.33e-2
extracellular space	GO:0005615	1.16e-1

observation where the D statistics of lognormal is smaller than gamma by order of magnitude. Note that the likelihood in tables 2 and 3 does not alway improve monotonically with the number of components. We believe this is due to the EM procedure getting trapped in poor quality local optima when more components are used.

Marginal Distribution of CSN3 Gene. CSN3 is a component of the COP9 signalosome complex, a complex involved in signal transduction. Figure 3 shows examples how one and two components mixtures of normal, lognormal, and gamma fit the marginal distribution of these gene. It is taken from GDS596 dataset.

Gene Ontology Comparison. From the two datasets we identified *two component* genes, the genes with probes for which the BIC criteria suggested two components for the learned lognormal mixture (there are 256 such genes for GDS596 and 56 for GDS592). We examined the overrepresented terms for each datasets using the web tool Babelomics (http://www.babelomics.org/) [14]. One common cellular component term - *intracellular* - that occurs in both of the datasets.

4   Conclusions

In this paper we provide a statistical framework using normal, lognormal, and gamma mixture models for analyzing the marginal distributions of expression levels in gene probes. To our knowledge this is the first study that provides such a framework for analyzing expression data.

Although theoretically gamma distributions are capable of modeling skewed distributions, our experiments showed that lognormal appears to be more suitable in modeling the marginal distribution of gene expression. We also showed that amongst the two model selection criteria we used, BIC is more accurate in selecting the number of components for lognormal and gamma mixtures. AIC on the other hand tends to over estimate the number of components.

We hypothesize that different functional categories of genes (*e.g.* transcription factors, kinases, structural proteins, etc) may show similar marginal distributions. Unfortunately this expectation is not clearly supported by our study. Only the single, vague gene ontology term *intracellular* was found to be over-represented in both datasets. We believe follow-up experiments are necessary to determine if this is a due to the quantity/quality of the expression data used, a deficiency in our methodology, or whether our hypothesis is simply wrong.

To achieve more definitive results we are now preparing to analyze a much larger dataset including multiple GEO datasets. This will be essential to sample the expression probes at the resolution needed to accurately model multimodal marginal distributions. Our results should provide some guidance in the development of informed priors or gene specific normalization for use with gene network inference.

References

1. Hoyle, D., Rattray, M., Jupp, R., Brass, A.: Making sense of microarray data distributions. Bioinformatics 18, 576–584 (2002)
2. Ji, Y., Wu, C., Liu, P., Wang, J., Coombes, K.R.: Applications of beta-mixture models in bioinformatics. Bioinformatics 21(9), 2118–2122 (2005)
3. Kuznetsov, V.: Family of skewed distributions associated with the gene expression and proteome evolution. Signal Process. 83(4), 889–910 (2003)
4. Mayrose, I., Friedman, N., Pupko, T.: A gamma mixture model better accounts for among site rate heterogeneity. Bioinformatics 21(2), 151–158 (2005)
5. Dennis, B., Patil, G.P.: The gamma distribution and weighted multimodal gamma distributions as models of population abundance. Mathematical Biosciences 68, 187–212 (1984)
6. Keles, S.: Mixture modeling for genome-wide localization of transcription factors. Biometrics 63(1), 2118–2122 (2007)
7. Limpert, E., Stahel, W., Abbt, M.: Log-normal distributions across the sciences: keys and clues. Bioscience 51(5), 341–352 (2001)
8. Konishi, T.: Parametric treatment of cDNA microarray data. Genome Informatics 7(13), 280–281 (2002)
9. Dempster, N.M., Laird, A.P., Rubin, D.B.: Maximum likelihood from incomplete data via the EM algorithm. J.R. Stat. Soc. 39(B), 1–38 (1977)
10. Akaike, H.: Information theory and extension of the maximum likelihood principle. In: Second International Symposium on Information Theory, pp. 267–281 (1973)
11. Schwarz, G.: Estimating the dimension of a model. The Annals of Statistics 6, 461–464 (1978)

12. Barrett, T., Troup, D.B., Wilhite, S.E., Ledoux, P., Rudnev, D., Evangelista, C., Kim, I.F., Soboleva, A., Tomashevsky, M., Edgar, R.: NCBI GEO: mining tens of millions of expression profiles - database and tools update. Nucleic Acids Research 35(Database-issue), 760–765 (2007)
13. Su, A.I., Wiltshire, T., Batalov, S., Lapp, H., Ching, K.A., Block, D., Zhang, J., Soden, R., Hayakawa, M., Kreiman, G., Cooke, M.P., Walker, J.R., Hogenesch, J.B.: A gene atlas of the mouse and human protein-encoding transcriptomes. Proc. Natl. Acad. Sci. USA 101(16), 6062–6067 (2004)
14. Al-Shahrour, F., Minguez, P., Vaquerizas, J., Conde, L., Dopazo, J.: Babelomics: a suite of web tools for functional annotation and analysis of groups of genes in high-throughput experiments. Nucleic Acids Research 33, W460 (2005)

Appendix: Density Functions

Let x be the expression value of a gene probe and j be the component's index. The density function for each component is denoted as $p(x|j)$. The density function for each type of distribution used in our mixture models is as follows:

1. Normal distribution.

$$p(x|j) = \frac{1}{\theta_{B_j}\sqrt{2\pi}}exp\left[-\frac{(x-\theta_{A_j})^2}{2\theta_{B_j}^2}\right]$$

2. Lognormal distribution.

$$p(x|j) = \frac{1}{x\theta_{B_j}\sqrt{2\pi}}exp\left[-\frac{(ln(x)-\theta_{A_j})^2}{2\theta_{B_j}^2}\right]$$

3. Gamma distribution.

$$p(x|j) = x^{\theta_{A_j}-1}\frac{e^{x/\theta_{B_j}}}{\theta_{B_j}^{\theta_{A_j}}G(\theta_{A_j})}$$

where $G(\theta_A) = 2\int_0^\infty e^{-t^2}t^{2\theta_A-1}\,dt$ is the Gamma function.

θ_A and θ_B correspond to the parameters in each distributions (i.e. *shape, scale* in gamma and *mean, standard deviation* in normal and lognormal distributions).

Combining Biochemical Features and Evolutionary Information for Predicting DNA-Binding Residues in Protein Sequences

Liangjiang Wang

Department of Genetics and Biochemistry, Clemson University, Clemson, SC 29634, USA
liangjw@clemson.edu

Abstract. This paper describes a new machine learning approach for prediction of DNA-binding residues from protein sequence data. Several biologically relevant features, including biochemical properties of amino acid residues and evolutionary information of protein sequences, were selected for input encoding. The evolutionary information was represented as position-specific scoring matrices (PSSMs) and several new descriptors developed in this study. The sequence-derived features were then used to train random forests (RFs), which could handle a large number of input variables and avoid model overfitting. The use of evolutionary information together with biochemical features was found to significantly improve classifier performance. The RF classifier was further evaluated using a separate test dataset. The results suggest that the RF-based approach gives rise to more accurate prediction of DNA-binding residues than previous studies.

Keywords: DNA-binding site prediction, feature extraction, evolutionary information, random forests, machine learning.

1 Introduction

Protein-DNA interactions are essential for many biological processes. For instance, transcription factors activate or repress downstream gene expression by binding to specific DNA motifs in promoters [1]. Protein-DNA interactions also play important roles in DNA replication, repair and modification. To understand the molecular mechanism of protein-DNA interactions, it is important to identify the DNA-binding residues in DNA-binding proteins. The identification can be straightforward if the structure of a protein-DNA complex is already known. However, it is rather expensive and time-consuming to solve the structure of a protein-DNA complex. Currently, only a few hundreds of protein-DNA complexes have structural data available in the Protein Data Bank [2]. With the rapid accumulation of sequence data from many genomes, computational methods are needed for predicting DNA-binding residues from protein sequence information. The prediction results may be used for gene functional annotation, protein-DNA docking and experimental studies such as site-directed mutagenesis.

T.-h. Kim et al. (Eds.): FGCN 2008 Workshops and Symposia, CCIS 28, pp. 176–189, 2009.
© Springer-Verlag Berlin Heidelberg 2009

Machine learning has recently been applied to sequence-based prediction of DNA-binding residues. The problem can be specified as follows: given the amino acid sequence of a protein that is supposed to interact with DNA, the task is to predict which amino acid residues may be located at the interaction interface. Both the structure of the protein and the sequence of the target DNA are assumed to be unknown. Although some experimental observations have been made for DNA-binding residues in protein structures, the molecular recognition mechanism is still poorly understood [3]. It is desired that machine learning methods can be used to model the complex patterns hidden in the available structural data and the resulting classifier can be applied to reliable identification of DNA-binding residues in protein sequences. Therefore, it is a challenging task to predict DNA-binding residues from amino acid properties and local sequence patterns.

Several studies have been reported for sequence-based prediction of DNA-binding residues. Ahmad et al. [4] analyzed the structural data of representative protein-DNA complexes, and used the amino acid sequences in these structures to train artificial neural networks (ANNs) for DNA-binding site prediction. Yan et al. [5] constructed Naïve Bayes classifiers using the amino acid identities of DNA-binding sites and their sequence neighbors (context information). However, the prediction accuracy was not high, probably because amino acid sequences were directly used for classifier construction in these studies.

Classifier performance has been shown to be enhanced by using evolutionary information for input encoding. Ahmad and Sarai [6] developed an ANN-based method to utilize evolutionary information in terms of position-specific scoring matrices (PSSMs). It was found that the average of sensitivity and specificity could be increased by up to 8.7% using PSSMs when compared with ANN predictors using sequence information only [4]. More recently, PSSMs were also used to train support vector machines (SVMs) and logistic regression models for accurate prediction of DNA-binding residues [7,8]. For a given protein sequence, its PSSM can be derived from the result of a PSI-BLAST search against a large sequence database. The scores in the PSSM indicate how well each amino acid position of the query sequence is conserved among its homologues. Since functional sites, including DNA-binding residues, tend to be conserved among homologous proteins, PSSM scores can provide relevant information for classifier construction. However, PSSM is rather designed for PSI-BLAST searches, and it may not contain all the evolutionary information for modeling DNA-binding sites.

In our previous studies [9,10], ANN and SVM classifiers were constructed using relevant biochemical features, including the hydrophobicity index, side chain pK_a value, and molecular mass of an amino acid. These features were used to represent biological knowledge, which might not be learned from the training data of DNA-binding residues. It was found that classifier performance was significantly improved by the use of biochemical features for input encoding, and the SVM classifier outperformed the ANN predictor. However, it is still unknown whether classifier performance can be further improved by combining the biochemical features with evolutionary information.

The main objective of this study is to improve classifier performance by combining different types of biological knowledge, including new descriptors of evolutionary information, PSSM and biochemical features. This approach gives rise to a large number of input variables. For DNA-binding site prediction, a data instance normally includes multiple neighboring residues for providing context information, and each residue is encoded using many feature values (*e.g.*, 20 PSSM scores for each residue). Considering the relatively small dataset currently available for modeling DNA-binding sites, too many input variables may result in model overfitting. To avoid this potential pitfall, classifiers have been constructed using the random forest (RF) learning algorithm, which has the capability to handle a large number of input variables and avoid model overfitting [11]. The results obtained in this study suggest that DNA-binding site prediction can be significantly improved by combining relevant biochemical features with several descriptors of evolutionary information for input encoding.

2 Methods

2.1 Data Preparation

Two amino acid sequence datasets, PDNA-62 and PDC25t, were derived from structural data of protein-DNA complexes available at the Protein Data Bank (http://www.rcsb.org/pdb/). The PDNA-62 dataset was used for training classifiers in this work as in the previous studies [6-10]. PDNA-62 was derived from 62 structures of representative protein-DNA complexes, and the amino acid sequences in this dataset shared less than 25% identity. The PDC25t dataset was derived from the protein-DNA complexes that were not included in PDNA-62. The sequences in PDC25t had less than 25% identity among them as well as with the sequences in PDNA-62. In this study, PDC25t was used as a separate test dataset for classifier performance evaluation and comparison.

DNA-binding residues in protein-DNA complexes were identified using atom distance or solvent accessible surface area (ASA). For the atom distance-based method, an amino acid residue was designated as a binding site if the side chain or backbone atoms of the residue fell within a cutoff distance of 3.5 Å from any atoms of the DNA molecule in the complex, and all the other residues were regarded as non-binding sites. This definition of DNA-binding residues was used in previous studies by us [9,10] as well as others [4,6-8]. It is noteworthy that both PDNA-62 and PDC25t are imbalanced datasets with ~15% residues labeled as DNA-binding and ~85% residues being non-binding.

In some previous studies [5,12], DNA-binding residues were also labeled by using the change of ASA during protein-DNA complex formation. A residue was assumed to be a binding site if the residue's ASA lost at least one square angstrom (ΔASA $\geq$ 1 Å^2) after the protein bound to DNA. In this study, two ASA values were computed for each amino acid residue in the protein-DNA complex or unbound protein (structural data with the DNA coordinates removed) using GETAREA [13]. The residue's ΔASA was then calculated by subtracting its ASA in the protein-DNA complex from that in the unbound protein. The above two definitions of DNA-binding residues were shown to give rise to slightly different datasets [9], and thus could affect classifier evaluation and comparison.

2.2 Training Strategy

Classifiers were trained using residue-wise data instances derived from the sequence dataset (PDNA-62). Each data instance consisted of eleven residues with the target residue positioned in the middle of the subsequence. From a protein sequence with n amino acid residues, a total of $(n - 10)$ data instances were extracted. A data instance was labeled as 1 (positive) if the target residue was DNA-binding or 0 (negative) if the target residue was non-binding. The context information provided by the five neighboring residues on each side of the target residue was previously shown to be optimal for sequence-based prediction of DNA-binding residues [9,10].

To generate the input vector, each residue was represented with three biochemical features and several descriptors of evolutionary information (see below). The three biochemical features, including the side chain pK_a value (feature K), hydrophobicity index (feature H) and molecular mass (feature M) of an amino acid, were previously demonstrated to be relevant for predicting DNA-binding residues [9,10].

2.3 Evolutionary Information

Considering the great complexity of protein-DNA interactions, the labeled datasets derived from the available structures are rather small in size. On the other hand, there are abundant unlabeled sequence data in public databases such as UniProt [14]. The unlabeled data contain evolutionary information about the conservation of each sequence position, and DNA-binding residues tend to be conserved among homologous proteins [15].

For a given protein sequence p, its homologues in a reference database can be retrieved and aligned to p using PSI-BLAST [16]. The sequence alignment is then used to compute evolutionary conservation scores for each residue in p. In this study, the protein sequence dataset UniProtKB (http://www.pir.uniprot.org/) was used as the reference database, and PSI-BLAST was run for three iterations with the E-value threshold set to 1e-5. The following descriptors of evolutionary information have been investigated in this study for predicting DNA-binding residues:

(1) BLAST-based conservation score (feature B): Let $H_p = \{h_1, h_2, ..., h_n\}$ be the set of n hits $(n > 0)$ in the PSI-BLAST search for a query sequence p. Each hit is a pair-wise sequence alignment, in which PSI-BLAST indicates whether two aligned residues are identical or show similarity based on the BLOSUM62 scoring matrix [16]. The B score for the residue a_i at position i in p is computed as follows:

$$B_{a_i}^{p} = \frac{\sum\limits_{h_j \in H_p} f(a_i, h_j)}{n + \dfrac{c}{n}} \tag{1}$$

where $f(a_i, h_j)$ is set to 1 if a_i is aligned to an identical or similar residue in h_j, or 0 otherwise, and c is a pseudo-count (set to 10 in this work). The term (c/n) is used to scale the feature value, and it becomes smaller when n gets larger. If p has no hit in the database ($n = 0$), the feature value is set to 0. The B score was used to construct artificial neural network classifiers in our previous study [9].

(2) Mean and standard deviation of biochemical feature values: For each residue a_i in the sequence p, the mean ($\overline{X}^p_{a_i}$) and standard deviation (σ) of a biochemical feature X, $X \in \{H, K, M\}$, are calculated as follows:

$$\overline{X}^p_{a_i} = \frac{\sum_{h_j \in H_p} \chi(a_i, h_j)}{n} \tag{2}$$

$$\sigma(X^p_{a_i}) = \sqrt{\frac{\sum_{h_j \in H_p}(\chi(a_i, h_j) - \overline{X}^p_{a_i})^2}{n-1}} \tag{3}$$

where $\chi(a_i, h_j)$ is the value of feature X for the amino acid residue in h_j, which is aligned to a_i at position i in p. The mean of feature X, also referred to as H_m, K_m or M_m in this paper, captures the biochemical properties of an amino acid position in the sequence alignment. It has been shown that basic and polar amino acids are overrepresented while acidic and hydrophobic amino acids are underrepresented in the population of DNA-binding sites [4,9]. The standard deviation of feature X, also called H_d, K_d or M_d, reveals how well the biochemical properties of an amino acid position are conserved in the aligned homologous sequences.

(3) Position-specific scoring matrix (PSSM): The PSSM scores are generated by PSI-BLAST [14], and there are 20 values for each sequence position. The evolutionary information captured by PSSMs was previously shown to improve the performance of artificial neural network and support vector machine classifiers for predicting DNA-binding residues [6,7]. Nevertheless, PSSM is rather designed for BLAST searches, and it may not contain all the evolutionary information for predicting DNA-binding residues, especially with regard to the relevant biochemical properties discussed above.

2.4 Random Forests

One potential problem with the use of evolutionary information is that the number of input variables becomes very large. In particular, the PSSM descriptor has 20 values for each sequence position. For a data instance with eleven residues, PSSM alone gives rise to 220 inputs. Considering the relatively small size of the training dataset, too many inputs can result in model overfitting. This problem may be solved using random forests [11].

Random forests (RFs) use a combination of independent decision trees to improve classification accuracy. Specifically, each decision tree in a forest is constructed using a bootstrap sample from the training data. During tree construction, m variables out of all the n input variables ($m \ll n$) are randomly selected at each node, and the tree node are split using the selected m variables. Because of the random feature selection, RFs can handle a large number of input variables and avoid overfitting. For classifying a data instance, a RF classifier combines the votes made by the decision trees, and gives the most popular class as the output of the ensemble. It has been shown that RFs outperform AdaBoost ensembles on noisy datasets, and can work well on data with many weak inputs [11]. These characteristics of RFs are appealing since the DNA-binding data appear to be noisy and may contain many weak sequence-derived features.

The software available at http://www.stat.berkeley.edu/~breiman/RandomForests/ was used to construct the RF classifiers in this study with the default parameter settings for training. In particular, the number of variables selected to split each node (m) was set to the floor of square root of the total number of input variables (default setting). Other values of m were also tested, but did not result in significant improvement of classifier performance.

2.5 Classifier Evaluation

A fivefold cross-validation approach was used to provide the initial estimation of classifier performance on the PDNA-62 dataset. The trained classifier was further evaluated using the PDC25t dataset. The following performance measures were used in this work:

$$Accuracy = \frac{TP + TN}{TP + TN + FP + FN} \tag{4}$$

$$Sensitivity = \frac{TP}{TP + FN} \tag{5}$$

$$Specificity = \frac{TN}{TN + FP} \tag{6}$$

$$Strength = \frac{Sensitivity + Specificity}{2} \tag{7}$$

where TP is the number of true positives (binding residues with positive predictions); TN is the number of true negatives (non-binding residues with negative predictions); FP is the number of false positives (non-binding residues but predicted as binding sites); and FN is the number of false negatives (binding residues but predicted as non-binding sites). Since the datasets used in this study are imbalanced, the overall accuracy alone could be misleading. Thus, both sensitivity and specificity are also computed from prediction results. Furthermore, the average of sensitivity and specificity, referred to as strength in this paper, may provide a fair measure of classifier performance as shown in previous studies [4,9].

Matthews Correlation Coefficient (MCC) is commonly used as a measure of the quality of binary classifications [17]. It measures the correlation between predictions and the actual class labels. However, for imbalanced datasets, different tradeoffs of sensitivity and specificity may give rise to different MCC values for a classifier. MCC is defined as:

$$MCC = \frac{TP \times TN - FP \times FN}{\sqrt{(TP + FP)(TP + FN)(TN + FP)(TN + FN)}} \tag{8}$$

The Receiver Operating Characteristic (ROC) curve is probably the most robust approach for classifier evaluation and comparison [18]. The ROC curve is drawn by plotting the true positive rate (*i.e.*, sensitivity) against the false positive rate, which equals to (1 – specificity). In this work, the ROC curve has been generated by varying the output threshold of a classifier and plotting the true positive rate against false positive rate for each threshold value. The area under the ROC curve (AUC) can be used as a reliable measure of classifier performance [19]. Since the ROC plot is a unit square, the maximum value of AUC is 1, which is achieved by a perfect classifier. Weak classifiers and random guessing have AUC values close to 0.5.

3 Results

3.1 Prediction of DNA-Binding Residues Using Random Forests

Random forests (RFs) were first trained with three biochemical features that were previously used to construct ANN and SVM predictors [9,10]. The biochemical features, including the hydrophobicity index (feature H), side chain pK_a value (K) and molecular mass (M) of an amino acid, were shown to provide relevant information for predicting DNA-binding residues [9]. The input vector contained 33 feature values because each data instance was a subsequence of eleven consecutive residues with the target residue in the middle position. The context information provided by the ten neighboring residues was found to be optimal for DNA-binding site prediction [9,10]. Table 1 shows the results obtained using the PDNA-62 dataset with the atom distance-based definition of DNA-binding residues (see Methods).

Table 1. Performance of different classifiers constructed using biochemical features

Classifier	Accuracy (%)	Sensitivity (%)	Specificity (%)	Strength (%)	MCC	ROC AUC
RF	70.23	73.46	69.68	71.57	0.32	0.78
SVM	70.31	69.40	70.47	69.94	0.29	0.75
ANN	64.38	71.33	63.51	67.42	0.27	0.73

The RF classifier constructed using the biochemical features achieved 70.23% overall accuracy with 73.46% sensitivity and 69.68% specificity in fivefold cross-validation experiments. Since the dataset was imbalanced with only 15% of the amino acid residues as DNA-binding sites, the performance of the RF classifier was also measured by the average of sensitivity and specificity (prediction strength = 71.57%), Matthews correlation coefficient (MCC = 0.32), and the area under the receiver operating characteristic curve (ROC AUC = 0.78). Different training parameters were tested for constructing the RF classifier, and the above performance measures were obtained using 1000 decision trees in the forest with $m = 5$. This classifier had the highest level of prediction strength. It should be noted that the MCC value was obtained using the given sensitivity and specificity. Higher MCC values might be obtained by using other tradeoffs of sensitivity and specificity for the RF classifier. In this study, the prediction strength was used for the initial selection of the best classifier, which was then evaluated using the other performance measures.

The results suggest that, with the three biochemical features, the RF classifier is slightly more accurate than the ANN and SVM predictors [9,10]. By using the same dataset (PDNA-62), the ANN and SVM predictors achieved the prediction strength of 67.42% and 69.94%, respectively. The ROC AUC and MCC of the ANN and SVM predictors are also slightly less than those of the RF classifier. However, RFs may have major advantages in handling a large number of input variables and avoiding model overfitting when various descriptors of evolutionary information are used for classifier construction.

3.2   Effect of Evolutionary Information on Classifier Performance

RF classifiers were constructed using different types of evolutionary information, including the BLAST-based conservation score, position-specific scoring matrices (PSSMs), and the means and standard deviations of biochemical feature values. For the first set of experiments, the PDNA-62 dataset with the atom distance-based definition of DNA-binding residues was used to construct RF classifiers. The conservation score (B) was previously used to train ANN classifiers for DNA-binding site prediction [9]. As shown in Table 2, the prediction strength (73.23%), MCC (0.34) and ROC AUC (0.81) are slightly improved by adding the B score to the three biochemical features (H, K, and M), suggesting that the conservation score does not capture most of the evolutionary information for sequence-based prediction of DNA-binding residues.

Consistent with previous studies [6-8], the use of PSSM scores for input encoding was found to significantly improve the classifier performance. As shown in Table 2, the RF classifier achieved 76.82% prediction strength with 79.26% sensitivity and 74.38% specificity. It also had higher MCC (0.40) and ROC AUC (0.85) than the RF classifier constructed using the three biochemical features alone (MCC = 0.32 and AUC = 0.78). The results were obtained using 1000 decision trees in the forest and the training parameter m set to 15 ($\lfloor\sqrt{253}\rfloor$). Because each residue was encoded with 20 PSSM scores and 3 biochemical features, the input vector contained 253 values for a data instance with eleven residues.

Table 2. Effect of evolutionary information on the performance of random forest classifiers constructed using the PDNA-62 dataset with the atom distance-based definition of DNA-binding residues

Evolutionary information	Accuracy (%)	Sensitivity (%)	Specificity (%)	Strength (%)	MCC	ROC AUC
None	70.23	73.46	69.68	71.57	0.32	0.78
B	72.74	73.92	72.54	73.23	0.34	0.81
PSSM	75.09	79.26	74.38	76.82	0.40	0.85
H_m, H_d, K_m, K_d	74.78	77.70	74.29	75.99	0.39	0.84
PSSM, H_m, H_d, K_m, K_d	78.20	78.06	78.22	78.14	0.43	0.86

The new descriptors of evolutionary information, including the mean and standard deviation of the three biochemical features, were also found to improve classifier performance. These descriptors indicate how well the biochemical properties of an amino acid position are conserved in the sequence alignment obtained from PSI-BLAST search. As shown in Table 2, the use of H_m, K_m, H_d, and K_d for input encoding gave rise to 75.99% prediction strength with 77.70% sensitivity and 74.29% specificity. This classifier achieved similar levels of MCC (0.39) and AUC (0.84) as the RF constructed using PSSM. The RF classifier was constructed using 1000 decision trees and $m = 8$. However, adding M_m and M_d to the input vector did not result in further improvement of classifier performance (data not shown).

To investigate whether classifier performance could be further improved by combining the different types of evolutionary information for input encoding, RF classifiers were constructed using PSSM, H_m, H_d, K_m and K_d in addition to the three biochemical features. Since the input vector had 297 variables (27 inputs for each of the eleven residues in a data instance), the training parameter m was set to $\lfloor \sqrt{297} \rfloor = 17$ for the forest with 1000 decision trees. As shown in Table 2, the resulting classifier achieved the highest level of prediction strength at 78.14% with 78.06% sensitivity and 78.22% specificity. This RF also had the highest level of MCC (0.43) and ROC AUC (0.86) among all the classifiers (Table 2). The results suggest that the new descriptors capture certain evolutionary information that is not contained in PSSM, and thus combining the different types of evolutionary information for input encoding gives rise to the most accurate classifier for DNA-binding site prediction.

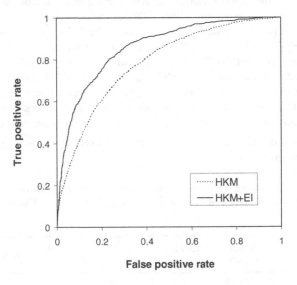

Fig. 1. ROC analysis to show the effect of evolutionary information on random forest classifiers constructed using the PDNA-62 dataset with the atom distance-based definition of DNA-binding residues. HKM represents the classifier trained with the three biochemical features (H, K and M), and HKM+EI indicates the classifier using two types of evolutionary information (PSSM, H_m, H_d, K_m and K_d).

Table 3. Effect of evolutionary information on the performance of random forest classifiers constructed using PDNA-62 with the ASA-based definition of DNA-binding residues

Evolutionary information	Accuracy (%)	Sensitivity (%)	Specificity (%)	Strength (%)	MCC	ROC AUC
None	70.64	70.40	70.71	70.55	0.35	0.77
B	72.44	73.09	72.26	72.67	0.39	0.81
PSSM	78.51	76.51	79.06	77.78	0.48	0.86
H_m, H_d, K_m, K_d	76.38	76.82	76.26	76.54	0.46	0.84
PSSM, H_m, H_d, K_m, K_d	78.55	77.92	78.72	78.32	0.49	0.87

The effect of evolutionary information on classifier performance has further been examined using ROC analysis. The ROC curves shown in Fig.1 have been generated by varying the output threshold of RF classifiers, and each point on a ROC curve represents a trade-off between sensitivity and specificity. For classifier performance comparison, the ROC curve of a more accurate classifier is closer to the left-hand and top borders of the plot. As shown in Fig. 1, the RF classifier trained with the two types of evolutionary information (HKM+EI) is clearly better than the classifier constructed using only biochemical features (HKM).

Next, we investigated the effect of evolutionary information on RF classifiers constructed using the PDNA-62 dataset with the ASA-based definition of DNA-binding residues. The ASA-based definition gave rise to more DNA-binding residues than the atom distance-based definition. While the ASA-based set of DNA-binding residues included 97.21% of the atom distance-based set of positive data instances (1,082 positive data instances), the ASA-based set also contained 553 positive data instances that were designated as non-binding residues by the atom distance-based definition. In other words, 33.82% of the DNA-binding residues defined by the ASA-based criterion were not included in the atom distance-based set of positive data instances.

As shown in Table 3, the classifier constructed without evolutionary information achieved 70.55% prediction strength with MCC = 0.35 and ROC AUC = 0.77, comparable to the levels of performance measures shown in Table 2. Adding the different descriptors of evolutionary information (B, PSSM, H_m, H_d, K_m and K_d) for input encoding improved the performance of RF classifiers. Furthermore, the best classifier in Table 3 was also obtained by combining the different descriptors of evolutionary information (PSSM, H_m, H_d, K_m and K_d) with the three biochemical features. This RF classifier had the prediction strength at 78.32% with 77.92% sensitivity and 78.72% specificity, MCC = 0.49 and ROC AUC = 0.87 (Table 3). The classifier performance improvement by using evolutionary information has further been confirmed in the ROC analysis (Fig. 2).

Therefore, with the ASA-based definition of DNA-binding residues, the use of evolutionary information was also found to significantly improve the performance of RF classifiers. In the previous study by Yan et al. [5], Naïve Bayes classifiers were constructed for predicting DNA-binding residues defined by the ASA-based criterion, and

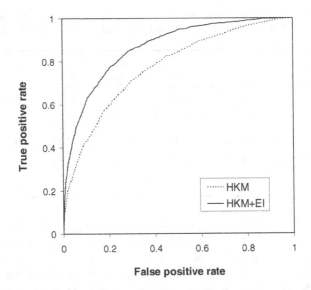

Fig. 2. ROC analysis to show the effect of evolutionary information on random forest classifiers constructed using the PDNA-62 dataset with the ASA-based definition of DNA-binding residues. HKM represents the classifier trained with the three biochemical features (H, K and M), and HKM+EI indicates the classifier using two types of evolutionary information (PSSM, H_m, H_d, K_m and K_d).

the best classifier trained with sequence identity and entropy achieved 78% overall accuracy but with only 41% sensitivity and MCC = 0.28. Although a different dataset was used for classifier construction and evaluation in the previous study [5], the RF classifier developed in the present study appears to be significantly more accurate than the Naïve Bayes classifier for DNA-binding site prediction. It is likely that the use of evolutionary information together with biochemical features for input encoding in this study but not in the previous study [5] is responsible for the improved classifier performance.

3.3 Classifier Evaluation Using a Separate Test Dataset

The results presented so far have been obtained from fivefold cross-validation experiments on the PDNA-62 dataset. To further evaluate the most accurate RF in Table 2 (also called BindN-RF), we prepared a separate test dataset (PDC25t), which shared less than 25% sequence identity with the PDNA-62 dataset. The RF classifier was also compared with two of the previously published classifiers (BindN and DBS-PSSM). BindN used the SVM classifier constructed using the three biochemical features in our previous study [10]. DBS-PSSM (http://www.netasa.org/dbs-pssm/) used the ANN predictor trained with PSSM and sequence information [6]. These two existing classifiers were chosen because they were constructed using the same training dataset (PDNA-62) as in the present study, and used the same distance-based criterion

Table 4. Performance comparison of different classifiers using a separate test dataset

Classifier	Accuracy (%)	Sensitivity (%)	Specificity (%)	Strength (%)	MCC	ROC AUC
BindN-RF	80.00	73.08	80.63	76.86	0.35	0.85
BindN	70.81	68.70	71.01	69.85	0.24	0.76
DBS-PSSM	67.91	37.48	70.72	54.10	0.05	0.55

to define DNA-binding residues. Other classifiers were constructed either using a different dataset (including some sequences in PDC25t) or with a different definition of DNA-binding residues.

As shown in Table 4, BindN-RF gives the best predictive performance with the prediction strength at 76.86% (73.08% sensitivity and 80.63% specificity), MCC = 0.35 and ROC AUC = 0.85. Importantly, the performance measures achieved by BindN-RF on the separate test dataset (PDC25t) are comparable with those from the fivefold cross-validation (Table 2), suggesting that overfitting has been avoided in the construction of the RF classifier. BindN is the second best classifier with the prediction strength at 69.85%, MCC = 0.24 and ROC AUC = 0.76. However, the ANN predictor trained with PSSM and sequence information (DBS-PSSM) shows very poor performance on the PDC25t dataset with only 54.10% prediction strength, MCC = 0.05 and ROC AUC = 0.55. The unexpected result for DBS-PSSM might be owing to poor generalization of the representative DNA-binding residues in the relatively small training dataset.

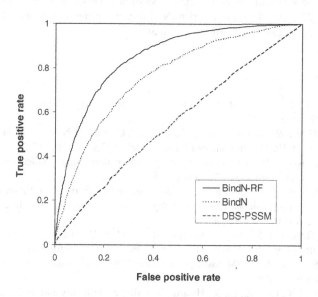

Fig. 3. ROC curves of three different classifiers (BindN-RF, BindN and DBS-PSSM) for sequence-based prediction of DNA-binding residues. The performance comparison is based on the PDC25t test dataset.

The ROC curves of the three classifiers (BindN-RF, BindN and DBS-PSSM) are shown in Fig. 3. Based on the predictions made for the PDC25t test dataset, the RF classifier (BindN-RF) clearly shows the best performance for almost all the trade-offs between sensitivity and specificity. The results suggest that the RF-based approach gives rise to more accurate prediction of DNA-binding residues in protein sequences than the previous methods.

4 Conclusions

Sequence-based prediction of DNA-binding residues can provide useful information for protein function annotation, protein-DNA docking and biological experiments. To improve the prediction accuracy, a random forest-based approach has been developed to combine relevant biochemical features with several descriptors of evolutionary information for input encoding. The new descriptors of evolutionary information have been shown to enhance classifier performance when they are used together with the biochemical features and position-specific scoring matrices. Thus, the new descriptors capture certain evolutionary information that is not contained in position-specific scoring matrices previously used for DNA-binding site prediction. It has also been shown in this study that evolutionary information can enhance classifier performance for predicting DNA-binding residues defined by both the atom distance-based and ASA-based criteria. The random forest-based approach gives rise to more accurate prediction of DNA-binding residues than previously published methods. By using a separate test dataset, the best random forest classifier achieved 80.00% overall accuracy with 73.08% sensitivity and 80.63% specificity. This classifier is currently being used to upgrade our web server, BindN (http://bioinfo.ggc.org/bindn/), which has been frequently accessed for biological research.

References

1. Ptashne, M.: Regulation of transcription: from lambda to eukaryotes. Trends Biochem. Sci. 30, 275–279 (2005)
2. Berman, H.M., Westbrook, J., Feng, Z., Gilliland, G., Bhat, T.N., Weissig, H., Shindyalov, I.N., Bourne, P.E.: The Protein Data Bank. Nucleic Acids Res. 28, 235–242 (2000)
3. Sarai, A., Kono, H.: Protein-DNA recognition patterns and predictions. Annu. Rev. Biophys. Biomol. Struct. 34, 379–398 (2005)
4. Ahmad, S., Gromiha, M.M., Sarai, A.: Analysis and prediction of DNA-binding proteins and their binding residues based on composition, sequence and structural information. Bioinformatics 20, 477–486 (2004)
5. Yan, C., Terribilini, M., Wu, F., Jernigan, R.L., Dobbs, D., Honavar, V.: Predicting DNA-binding sites of proteins from amino acid sequence. BMC Bioinformatics 7, 262 (2006)
6. Ahmad, S., Sarai, A.: PSSM-based prediction of DNA binding sites in proteins. BMC Bioinformatics 6, 33 (2005)
7. Kuznetsov, I.B., Gou, Z., Li, R., Hwang, S.: Using evolutionary and structural information to predict DNA-binding sites on DNA-binding proteins. Proteins 64, 19–27 (2006)
8. Hwang, S., Gou, Z., Kuznetsov, I.B.: DP-Bind: a web server for sequence-based prediction of DNA-binding residues in DNA-binding proteins. Bioinformatics 23, 634–636 (2007)

9. Wang, L., Brown, S.J.: Prediction of DNA-binding residues from sequence features. J. Bioinform. Comput. Biol. 4, 1141–1158 (2006)
10. Wang, L., Brown, S.J.: BindN: a web-based tool for efficient prediction of DNA and RNA binding sites in amino acid sequences. Nucleic Acids Res. 34, W243–W248 (2006)
11. Breiman, L.: Random forests. Machine Learning 45, 5–32 (2001)
12. Jones, S., Shanahan, H.P., Berman, H.M., Thornton, J.M.: Using electrostatic potentials to predict DNA-binding sites on DNA-binding proteins. Nucleic Acids Res. 31, 7189–7198 (2003)
13. Fraczkiewicz, R., Braun, W.: Exact and efficient analytical calculation of the accessible surface areas and their gradients for macromolecules. J. Comp. Chem. 19, 319–333 (1998)
14. Wu, C.H., Apweiler, R., Bairoch, A., Natale, D.A., Barker, W.C., Boeckmann, B., Ferro, S., Gasteiger, E., Huang, H., Lopez, R., Magrane, M., Martin, M.J., Mazumder, R., O'Donovan, C., Redaschi, N., Suzek, B.: The Universal Protein Resource (UniProt): an expanding universe of protein information. Nucleic Acids Res. 34, D187–D191 (2006)
15. Panchenko, A.R., Kondrashov, F., Bryant, S.: Prediction of functional sites by analysis of sequence and structure conservation. Protein Sci. 13, 884–892 (2004)
16. Altschul, S.F., Madden, T.L., Schaffer, A.A., Zhang, J., Zhang, Z., Miller, W., Lipman, D.J.: Gapped BLAST and PSI-BLAST: a new generation of protein database search programs. Nucleic Acids Res. 25, 3389–3402 (1997)
17. Baldi, P., Brunak, S., Chauvin, Y., Andersen, C.A.F., Nielsen, H.: Assessing the accuracy of prediction algorithms for classification: an overview. Bioinformatics 16, 412–424 (2000)
18. Swets, J.A.: Measuring the accuracy of diagnostic systems. Science 240, 1285–1293 (1988)
19. Bradley, A.P.: The use of the area under the ROC curve in the evaluation of machine learning algorithms. Pattern Recognition 30, 1145–1159 (1997)

A Predictive Approach for Selecting Suitable Computing Nodes in Grid Environment by Using Data Mining Technique

Asgarali Bouyer and Seyed Mojtaba Mousavi

Islamic Azad University- Miyandoab Branch,
Miyandoab Iran
{basgarali2,Mmseyed2}@Siswa.utm.my

Abstract. Grid Computing is a new technology for developing resources sharing such as computational resources, data and so on between many organizations in different geographical locations. Therefore resource management for finding, selecting, and scheduling is more important in the grid. This paper has concentrated on nodes selection to assign proper jobs. There are many methods to do this work but each of them has weakness. In this paper, we propose a new approach for fair Grid Resource Broker (GRB) to select optimal nodes, and then we compare it to other existing methods. Our proposed approach applies Fuzzy decision tree as evolution method for finding suitable resources in existing pool of resources in the grid. Moreover, if our approach is not encountered with the lack of resources, it will acts in best form and assigns to each task enough computational power for it to finish within its deadline. The role of this resource broker service is executing of Fuzzy Decision Tree algorithm to find the best nodes according to the requirements and conditions of the job on the Grid. The results of experiments show a powerful effect in improving resource finding cycle.

Keywords: Grid Resource broker, Resource Selection, Data mining, Fuzzy Decision Tree.

1 Introduction

Grid is a decentralized heterogeneous system that made up virtual organizations (VOs). Each VO is composed of several different nodes. Each node can be server computers, desktop PCs, clusters, and other kinds of hardware, which are sharing some resources with other nodes. A main goal of grid computing is enabling applications to identify resources dynamically to create distributed computing environments [1].

In order to perform job scheduling and resource management at Grid level, usually it is used a Resource Broker or a meta-scheduler. A resource broker is fundamental in any large-scale Grid environment. The task of a Grid resource broker and scheduler is to dynamically identify and characterize the available resources, and to select and allocate the most proper resources for a given job. In a broker-based management system, brokers are responsible for finding and selecting best nods, ensuring the trustworthiness of the service provider. Resource selection is an important issue in a

T.-h. Kim et al. (Eds.): FGCN 2008 Workshops and Symposia, CCIS 28, pp. 190–205, 2009.

grid environment where a consumer and a service provider are distributed geographically across multiple administrative domains. Choosing the suitable resource for a user job to meet predefined constraints such as deadline, speedup and cost of execution is an important problem in grids. In our approach, we highly have solved some of these problems [2]. In general, a Grid broker must make resource selection decisions in an environment where it has no control over the local resources. The resources are distributed, and information about the resources is often limited or dated.

Many brokers use the GRIS (Grid Resource Information Service) in discovery and resource selection phase. The Grid Resource Information Service (GRIS) has been considered as a tool of inquiring resources on a computational grid for their current configuration, capabilities, and status. The GRIS is a distributed information service that can answer queries about a particular resource by directing the query to an information provider deployed as part of the Globus services on a grid resource. Using GRIS permanently cannot be reliable because all data about resource and its performance are not exist or maybe old. In proposed approach we use a local database to record any things and events about submitted tasks (job) and at the next time this information (in local DB) can be very helpful to obtain a background about past resource's treatment in order to prediction or another operations. In this paper we will not do a resource discovery method, but in fact we present a novel way for selecting the best nodes in pool of discovered nodes. Resource selection involves a set of factors including application execution time, available main memory, disk (secondary memory), resource access policies, etc. resource selection must consider information about resource reliability, prediction error probability, and real time execution. However, these various performance measures can be considered under the condition that the middleware allows adaptation of its internal scheduling with desired application's services. We have considered all of these factors in our approach. Also to reach for better selection we used the Decision Tree with Fuzzy Logic theory [3]. Induced decision trees are an extensively-researched solution to classification tasks but general Decision Tree have some weakness that they covered by Fuzzy Decision Tree (FDT).

The rest of this paper is organized as follows. Section 2 refers to previous research on resource brokering and scheduling. Section 3 describes Fuzzy Decision Tree Algorithm in our method and also we have described why FDT was used in this paper. In section 4 we will discusses on proposed architecture and provided applications for Grid resource broker, respectively. The experimental results and performance evaluation for our method has mentioned in section 5. Finally, section 6 concludes the paper.

2   Related Works

Many projects, such as DI-GRUBER [5], eNANOS [6], AppLes [7] and OGSI- based broker [4] have been performed on grid. In this section we introduce some of these brokers.

DI-GRUBER [5], an extension to the GRUBER brokering framework, was developed as a distributed grid USLA based resource broker that allows multiple decision points to coexist and cooperate in real-time. GRUBER has been implemented in both Globus Toolkit4 (GT4) and Globus Toolkit3 (GT3). The part of DI-GRUBER that dosing resource finding and selecting is called The GRUBER engine. GRUBER engine is the main component of the GRUBER architecture and that implements various algorithms to detect available resources and maintains a generic view of resource utilization in the grid [5]. The proposed mechanism allows resources at individual sites to be shared among multiple user communities. It is a distributed Grid brokering service that it addresses issues regarding how USLAs can be stored, retrieved, and disseminated efficiently in a large distributed environment. DI-GRUBER has some defect in resource matching to tasks and repartition [17], but other recommendations have been offered to solve its problem.

The eNANOS Resource Broker is an OGSI-Compliant resource broker developed as a Grid Service and is supported by Globus Toolkit middleware [6]. This Broker provides a set of Grid Service interfaces and a Java API which can be used from command-line clients, applications or portals. The eNANOS uses from the Grid Information Service (GIS) to obtain needed information. As you know, GIS has some problems such as data restriction, the timeworn data, updateability problem and so forth. eNANOS architecture also neither uses data mining methods to select the best nodes from the pool of discovered nodes, nor implements in Web Services (WS) bases frameworks.

The AppLes (Application Level Scheduling) focuses on developing scheduling agents for individual Grid applications [7]. AppLes agents have an application oriented scheduling mechanism, and use static or dynamic application and resource information to select a set of resources. However, they perform resource discovering and scheduling without considering resource owner policies. Also they do not support system-oriented or extensible scheduling policies.

Another resource broker service has been presented by Young-Seok Kim and et al. [4]. It is an OGSI- based broker that is supported by GT3. It is a new general purpose OGSI-compliant Grid resource broker service that performs resource discovering and scheduling with close interactions with GT3 Core and Base Services. This resource broker service considers resource owner policies as well as user requirements on the resources.

The EZ-Grid project [8] tries to develop a transparent view of Grid resources and a simplified job submission through the EZ-Grid resource broker. The system is developed in Java, using the Java CoG Kit. The aim of the system is to relieve the user from the complexity involved in making resource selections, job specifications and submissions while maintaining transparency in providing middleware services and best resource choices. The system provides an easy-to-use interface that allows for user authentication, application profiling, remote resource information display, job submission and job monitoring.

Another works have been done in resource selection field (e.g. Condor/G [15], Nimrod/G [18], and so forth), but we cannot introduce all of them in this paper.

Finally, we mention that none of those systems or brokers uses machine learning methods to find (select) the best nodes for purposed jobs.

3 Fuzzy Decision Tree

Decision Tree (DT) is one of the most popular methods for learning and reasoning from feature-based examples. Due to following causes DT is better for our work:

- Decision trees are simple to understand and interpret: At each internal node, a decision is based upon just one predictor variable and this makes it easy to follow. For example, to explain particular classification one need only look at the series of simple decisions that led to it. The final tree model can in-fact be cast into a set of rules one can follow to classify a given case.
- Decision tree methods have a built-in feature selection method that makes them immune to the presence of useless variables.
- Tree models are very adept at revealing complex interactions between variables. Each branch of a tree can contain different combinations of variables and the same variable can appear more than once in different parts of the tree.
- Is robust, perform well with large data in a short time. Large amounts of data can be analyzed using personal computers in a time short enough to enable stakeholders to take decisions based on its analysis.
- Is a white box model. If a given situation is observable in a model the explanation for the condition is easily explained by Boolean logic. An example of a black box model is an artificial neural network since the explanation for the results is excessively complex to be comprehended.
- It is possible to validate a model using statistical tests. That makes it possible to account for the reliability of the model.
- Data preparation for a decision tree is basic or unnecessary. Other techniques often require data normalization, dummy variables need to be created and blank values to be removed.
- And other advantage.

However general decision tree always has a deterministic result, and therefore this feature is not good in some application. Fuzzy decision Tree (FDT) is the generalization of decision tree in fuzzy environment. The knowledge represented by fuzzy decision tree is closer to the human classification [10]. FDT is combining symbolic decision trees with approximate reasoning offered by fuzzy representation. The intent is to exploit complementary advantages of both: popularity in applications to learning from examples and high knowledge comprehensibility of decision trees, ability to deal with inexact and uncertain information of fuzzy representation. FDT has some advantage as following [19].

1. FDT is lower than DT in train accuracy while higher in test accuracy clearly, which means that the generalization ability of FDT is better and FDT can describe the character of customer data better.
2. The tree build by FDT is smaller and the leaf number is less, which means that created rule is easy to understand.
3. The speed of FDT is more rapid than DT, which fits to handle huge database, such as customer database in corporations.

In our approach we used a Fuzzy decision tree (FDT). Fuzzy decision trees provide away to manipulate fuzzy information and continuous input/output models while maintaining the interpretability and effectiveness of classical decision trees.

3.1 Fuzzy Logic (FL)

Essentially, Fuzzy Logic (FL) is a multi-valued logic that allows middle values to be defined between conventional evaluations like yes/no, true/false, black/white, etc. Fuzzy Logic was introduced in 1965 by Prof. L. Zadeh at the University of California, Berkeley [9]. The basic notion of fuzzy systems is a fuzzy set. for example, to classify the fuzzy set of climate, which may be consisted of members like "Very cold", "Cold", "Warm", "Hot", and "Very hot". The theory of fuzzy sets enables us to structure and describe activities and observations, which differ from each other only vaguely, to formulate them in models and to use these models for various purposes - such as problem-solving and decision-making [9]. Suppose that $\mu S(x)$ (or $\mu(S, x)$) is the degree of membership of x in set S that $0 \leq \mu S(x) \leq 1$

$\mu S(x) = 0$ x is not at all in S,

$\mu S(x) = 1$ x is fully in S,

If $\mu S(x) = 0$ or 1, then the set S is crisp.

For example, pay attention to the diagram 1 (Fig. 1).

What is the meaning of 75 in diagram? We analyze this question as following:

A node that finished successfully 75% of its submitted jobs has simultaneously Low, Medium, and High efficiency in various degrees. For example, it can be interpret as 0.2 Low efficiency, 0.5 Medium efficiency and 0.3 High efficiency. In other word, there is much status that needs to use of Fuzzy logic or fuzzy algorithm. For instance, consider the following scenario:

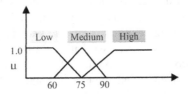

Fig. 1. This diagram shows the Node's efficiency

Let's assume that we have to select 3 computing nodes in between 20 existing nodes. At the first time, efficiency and availability for all nodes is evaluated. Suppose that 70% of nodes have efficiency in range 85 to 90, and 20% have near to 95 and 10% under 75. Therefore, in this case, the range (85-90) is considered as medium efficiency and if there are nodes with high efficiency, it is not needed to use medium efficiency nodes.

We will not discuss fuzzy set such natural extensions here and more about fuzzy logic can be found in [13].

3.2 Fuzzy Decision Tree Algorithm

This algorithm is a developed version of ID3 that operate on fuzzy set and it will produce a fuzzy decision tree (FDT). Before this, other researchers [3, 12] considered the FDT in their applications. Thus, their results showed that this algorithm is suitable for our approach. There are two important points in making and applying FDT [11]:

- Select the best attribute in each node to develop the tree: there are many criteria for this aim, but we will use one of them.
- Inference procedure from FDT. In the classification step for a new sample in FDT, we may encounter many leaf nodes with deferent confidence that offer some classes for purposed sample. Thus, the fitness mechanism selection is important here. Before we express the algorithm, we will consider some assumptions and notation:
- The training examples will be called E set with N example. Each example has N properties and every property A_j contain m_j linguistic term and so the number of output class will be as following.

$$E = \{e_1, e_2, ..., e_n\} \quad A = \{A_1, A_2, ..., A_N\} \quad \text{for } A_j; 1 \le j \le N \quad C = \{c_1, c_2, ..., c_k\}$$

$$\text{Fuzzy terms for } A_j \rightarrow \{V_{j1}, V_{j2}, ..., V_{jm_j}\}$$

- The set of exist examples in t nodes show by X.
- $\mu_{c_k}(x)$: represent the degree membership of example x belongs to the class c_k.
- $\mu_{v_{jL}}(x)$: represent the degree membership of crisp value for attribute j in example x belongs to the fuzzy term in j attribute. Also consider four following formulas:

$$P*(c_k) = \sum_{x \in X} \mu_{c_k}(x) \tag{1}$$

$$P*(v_k) = \sum_{x \in X} \mu_{v_L}(x) \tag{2}$$

$$P*(c_k / v_L) = \frac{P*(c_k \cap v_L)}{P*(c_k)} \tag{3}$$

$$P*(c_k \cap v_L) = \sum_{x \in X} \mu_{c_k}(x).\mu_{v_L}(x) \tag{4}$$

Creating a Fuzzy Decision Tree.
Step1: Start with all the training examples, having the original weights (degree membership of each sample to desired class is considered 1 value), in the root node. In other words, all training examples are used with their initial weights (This initial weight is not necessarily 1).
Step2: if in one of the node t with fuzzy set X one of the below condition is true, that node will consider as a leaf node.

Con1: for all examples of set X, the proportion for degree membership in a class to sum of degree membership of all data to different classes is equal or greater than θ_r.

$$[\sum_{x \in X} \mu_{c_k}(x) / \sum_{k=1}^{K} \sum_{x \in X} \mu_{c_k}(x)] \ge \theta_r \tag{5}$$

Con2: sum of degree membership of all data in set X, less than Threshold θr.

$$\sum_{k=1}^{K} \sum_{x \in X} \mu_{c_k}(x)] \le \theta_r \tag{6}$$

Con3: there have not been existed another attribute for selection.

Step3: if any conditions of step 2 for desired node are not true, then this node should be developed. Thus:

- *Step3.1:* find all attributes in a path from root node to desired node, and then remove it from attribute set. So remaining attribute will be more luck for selection.
- *Step3.2:* for every remaining attribute (A_i), we should select an attribute according to Entropy measure [10] to develop the tree (A_{max}).
- *Step3.3:* split X set into subsets X_1, X_2,..., X_{mAmax} so that, all elements in X_i, there is a coefficient of fuzzy term v_i for A_{max}.
- *Step3.4:* for every of these subsets, we will define nodes t_1, t_2,..., t_{mAmax} and then the edges are labeled by v_i values (i=1,2,..., m_{Amax}). Then, the degree membership for each example to new node will be computed as following.

$$\mu_{c_k}(x \in X_{v_i}) = \mu_{c_k}(x \in X).\mu_{Vi}(x \in X) \qquad (7)$$

- *Step3.5:* exchange each X_i with X and then repeat step 2 and 3.

4   System Architecture

This section describes the design of our approach as a part of Resource Broker, which is enhanced by adding the functionality to enable adaptability. We have shown a general architecture for this approach in figure 2 (Fig 2). The Resource Broker performs a number of basic functions. The first step is the discovery and selection of resources that best fit the needs of the Grid application. The broker will then submit jobs in the application to the chosen nodes. GRB use machine learning technique for Node selection. The broker thus handles submission of jobs but not how the job is actually executed on the resource. These actions are referred to as scheduling in the Resource Broker, but in this paper we will not discourse in this topic. Once jobs are being executed (or waiting to be executed), the broker monitors the resources and the progression of the jobs. Our application can change its behavior depending on available resources, optimizing itself to its dynamic environment for each new arrival job.

Our supplied application is performed on top of GT3. But it can be applied for GT4. For the nonce, we have provided an isolated application that can be worked based on GT3, for this purpose. The Result of every node is sent in an XML document and is stored in a Temporary XML Database (TXD).

4.1 Miner Application

To do this, we want to install a Miner Application (MA) for every node in a purposed grid. MA contains an internal small database (in log file role). One of the primary tasks of MA is writing log file. When desired node is connected to grid, MA must update its log file (insert a new record to database) or when a new job is submitted to this node, MA will update the related record (connection record) and also insert a new record for submitted job in desired table (contain job properties and status). This is more important because we want to know the number of submitted jobs that are successfully executed or failed on this node. At the moment, if the job is finished successfully or if the job is failed for any reason, thus, MA will update the database file (there is a Boolean field in table that if it is set to TRUE, this means that the related job has been finished successfully, otherwise, it means that the job is not successfully done and has failed). Also, we have considered some new tasks for Grid Resource Broker (GRB), which we have called Optimal GRB. Before selecting any nodes (for aimed job) by GRB, one of these tasks will be executed, this is responsible for sending a packet to each node on grid besides previous tasks. Needless to say, this task can be executed during recourses discovery operation by GRB. Further, as already stated, there are many different methods to find resources (nodes), but will not concentrate on how we can discovery nodes; and we will not mention them in this paper. Suppose that, there are many different nodes in our grid that are ready for executing jobs and we want to select some nodes in the pool of these nodes. At the beginning, GRB has sent a packet to each connected nodes to our grid. This packet contains some information about a new job (e.g. IP Sender, Size of the job, Size of needed RAM and HDD, Estimation time needed for execution, approximate execution start time, minimum power needs to CPU, OS status, etc.).

On the other side, when MA in node gets this packet, it will open the packet for analysis. If there are sufficient conditions (internal resources node) to do the desired job, MA will perform a data processing technique on its own mini-database (or its log file) to obtain some computation for this job. In many conditions if broker send a Data mining request to MA, the MA application will execute a DT algorithm and then it will send useful rules to broker. Some of produced results are as follows:

- Average Hit Ratio (AHR): This attribute represents an average rate of success in all previous times.
- Number of all submitted jobs on this node (AAJ).
- Number of all jobs submitted at this time, on the previous days, on this node (AATPJ).
- Number of all jobs successfully finished at this time, on the previous days (NSTP).
- Hit Ratio for this time-period on previous days (HRTP). For example, how many jobs in 1.30 AM o'clock to 2.00 o'clock have been executed?
- Average Size of successfully finished jobs (ASF).
- Average Response Time for finished jobs (ART).
- Average Response Time for jobs that have the same size as the purposed job and have been successfully finished (ARTSS).
- Hit Ratio for the last twenty jobs (HRT).
- Date, Time and Size of the last successfully finished job (LSJ).
- Date, Time and Size of the last failed job (LFJ).
- Size of the largest successfully finished job (LSI).
- Numbers of all previous jobs that almost have the same size as the purposed job (ASS). Needless to say, the size of the previous jobs is not exactly the same as the size of the desired jobs. For example, for a job with size=340KB we must find all of the previous jobs between 1K to 500KB size.
- Number of all previous jobs that have the same size as the purposed job and are successfully finished (NSS).
- Moreover, processor speed and CPU availability (Idleness) are important for choosing a node.

In addition to the node information, these results will be sent to GRB from any node. There, GRB will analyze them to select/deselect the desired nodes. We mention that always the last collected result will be saved by GRB.

4.2 Broker Layer

In this layer we have added two new sections beside general broker's sections. The first section is related to Request Broker section. This section must broadcast packet to all of the nodes in grid, then it must receive and save the sent results from each node in temporary XML database (TXD). Next, Recourse selector section will execute a Fuzzy decision Tree Algorithms on TXD (gathered result). We are doing this task in sub-section inside Resource Selector that we call FDT executer. Whenever this algorithm has finished its task, the next sub-section, SNJ (Selecting node for job), will use the result of the algorithm to identify suitable nodes.

FDT Executer. This section is considered for executing FDT algorithm on TXD data. As you know, FDT is a machine learning technique for extracting knowledge that is nearer human decision. In this research, we have used FDT algorithm (FID3), because it is reliable and flexible and also has a high accuracy in selecting samples. All used samples for both training and testing are extracted from the provided database (TXD). After that FDT algorithm was performed by FDT executer, therefore we can select a desired class for purposed jobs. Also, Jobs can be divided in several groups: high reliability jobs, real-time jobs, normal jobs, testing jobs and etc.

SNJ sub-layer
1. Based on the gathered results from FDT executer, this section will select appropriate nodes based on job conditions. There are many parameters in this section, but the main parameters that must be considered, are as follows:
2. Very High Reliability jobs : if we want to execute the desired job successfully with high reliability (response time is not very important), the AHR, HRTP, ASF, HRT measures are very important. There is a priority between these measures. For example, to achieve high reliability, AHR and then HRTP have a high priority. Of course, other measures are also important. SNJ will analyze these measures form gathered results (provided by FDT executer). For example, if there are six nodes that have almost the same AHR and HRTP, or ASF and so on, then other measures (e.g. ART or HRT) will select to evaluate the performance of these nodes. It is possible that there are some states in that SNJ cannot select its own nodes without limit. For example, suppose that SNJ needs to select seven nodes for doing the desired tasks, and there exist only five nodes with high reliability (AHR and HRTP over 95%) and also, if there exist other nodes with low reliability (less than 50%), then GRB can use other parameters to decreasing risk. For example, for two remaining nodes, SNJ can consider HRT parameter, because this is better than other Random-based methods. All of this will be done by SNJ. Also it can use multiversioning in hierarchical architecture to increase reliability [14]. In other words, it tries to start the versions through candidate nodes in parallel and distributed form

by dispatching some replicas of an offered job to the best-selected nodes with a special order.

3. Execution in Real Time: if we want to execute a job in real time status, so the CPU speed and ART have highest priority and next priority respectively belong to ARTSS, HRTP, AHR, ASF, and LSI and so on. Also processor's power and communication line bandwidth are important. In this approach we have concentrated on two kinds of jobs that are mentioned in this section.

For a fuzzy set, the idea of vagueness is introduced by assigning an indicator function that may take on values in the range 0 to 1. The following observations are considered:

- $Count(S_i)$: returned the number of successfully finished jobs on $node_i$
- $Count(ST_i)$: returned the number of successfully finished jobs in the last 20 submitted jobs on $node_i$
- $Count(AAJ_i)$: returned the number of all submitted jobs on $node_i$
- $Min(ART)$: return the minimum ART in between of all nodes
- $MAX(ASF)$: return the maximum ASF in between all nodes
- $Min(CPU\_SP_i)$= return the minimum CPU speed in between all nodes

Suppose that $1 \leq i \leq n$ (n is showing the number of nodes), here we mention how to compute or convert deterministic values to fuzzy sets. The most important attributes have been computed below (member functions) and they are very important to decide on selecting nodes:

$$\bullet \ A_{1i} = M(AHR_i) = \frac{Count(S_i)}{Count(AAJ_i)}$$

$$\bullet \ A2_i = M(ART_i) = 1 - \frac{(ART_i - Min(ART)}{ART_i}$$

$$\bullet \ A_{3i} = M(HRT_i) = \frac{Count(ST_i)}{20}$$

$$\bullet \ A_{4i} = M(HRTP_i) = \frac{(Count(NSTP_i) * Count(AAJ_i))}{Count(NSS_i)/Count(AAJ_i) - Count(AATPJ_i)^2 + 1}$$

$$\bullet \ A_{5i} = M(HRS_i) = \frac{Count(NSS_i)}{Count(ASS_i)}$$

$$\bullet \ A_{6i} = M(ASF_i) = \frac{ASF_i}{MAX(ASF)}$$

$$\bullet \ A_{7i} = M(speed_i) = 1 - \frac{Min(CPU\_SP)}{CPU\_SP_i}$$

$$\bullet \ A_{8i} = M(IDLE_i) = \frac{CPU \ IDLE_i}{100}$$

$$\bullet \ A_{9i} = M(RAM_i) = \frac{(free \ available \ RAM \ for \ node_i)}{Maximum \ Free \ avabilable \ RAM \ among \ all \ participating \ nodes}$$

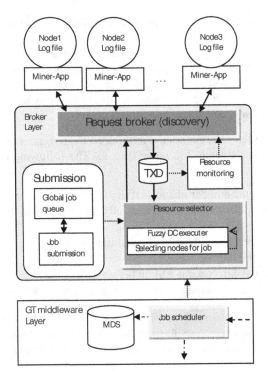

Fig. 2. General architecture for proposed approach

As you see, The A_5 shows the ratio of successful jobs that have similar size with the desired job to all successful finished jobs. A_7 shows the CPU power and A_8 shows the measure of system Idle in fuzzy range.

For the nonce, these nine attributes will be evaluated in fuzzy behavior. We must mention that based on the type of jobs, they will take a weight. This weight has been allocated based on empiric and the effect of each attributed in classification by DT. These weights are representing in Table 1. Now, to find a node with very high reliability rather than other nodes, we should compute the following computation (formula 8) for each node and then we will select that node with maximum Value according to formula 9. We will have this similar method for other type of jobs.

$$Val_{node_i} = \sum_{i=1}^{9}(WH_i * A_i) \qquad (8)$$

$$Selected\_node = \text{a node with Max}(\text{val}_{node_i}) \qquad (9)$$

5   Experimental Results and Discussion

We have designed two applications for our approach. The first application is executed on nodes (MA). In our experiments, eight resource computing nodes and one server are used to evaluate performance of this approach. The hardware information has been described in Table 2. These nodes communicate with server via internet. Then, the *MA application* is installed on nodes and *broker provider application* is installed on the server computer. We divide 24 hour in to following parts:

7-9	9-12	12-15	15-17	17-20	20-22	22-24	24-2	2-7

Table 1. Assign a weight for each attribute

Name of attributes	Weight for High reliability	Weight for Real-time	Weight for Normal jobs
$A_1 \rightarrow$	$WH_1=1$	$WR_1=0.7$	$WN_1=0.8$
$A_2 \rightarrow$	$WH_2=0.4$	$WR_2=1$	$WN_2=0.8$
$A_3 \rightarrow$	$WH_3=0.7$	$WR_3=0.6$	$WN_3=0.6$
$A_4 \rightarrow$	$WH_4=0.9$	$WR_4=0.4$	$WN_4=0.6$
$A_5 \rightarrow$	$WH_5=0.5$	$WR_5=0.2$	$WN_5=0.4$
$A_6 \rightarrow$	$WH_6=0.3$	$WR_6=0.1$	$WN_6=0.2$
$A_7 \rightarrow$	$WH_7=0.5$	$WH_7=0.9$	$WH_7=0.5$
$A_8 \rightarrow$	$WH_8=0.4$	$WH_8=0.8$	$WH_8=0.5$
$A_9 \rightarrow$	$WH_9=0.1$	$WH_9=0.2$	$WH_9=0.1$

Table 2. Hardware information

Name	Type of hardware
$Node_1$	Pentium4(Cache1MB),CPU2.2(INTEL), RAM 256
$Node_2$	Pentium(Cache2MB),CPU2.4(INTEL), RAM 512
$Node_3$	INTEL Pentium,CPU3.0(GLI),2, RAM 1G
$Node_4$	Intel(R) Core(TM)2 Duo CPU 2.16GHz RAM(3.49 GB)
$Node_5$	Intel(R) Core(TM)2 Duo CPU 2.16GHz RAM(3.49 GB)
$Node_6$	Intel(R) Core(TM)2 Duo CPU 2.16GHz RAM(3.49 GB)
$Node_7$	Intel(R) Core(TM)2 Duo CPU 2.16GHz RAM(2.9 GB)
$Node_8$	HP ProLiant ML370 G4 High Performance – Intel Xeon 3.4 GHz (2 processors)L2 cache(RAM 8G)
server	Pentium4(Cache2MB),CPU3.0(INTEL), RAM 1G

In the first six days we have used MA Application but we didn't use the result of MA in our broker application. Moreover we always have sent a job for all available nodes. After that, we have activated broker provider to select only suitable nodes. Therefore, in seventh day, we have taken the listed results in Table 3 from available nodes in the morning in order to execution a high reliability job with size 4.47 MB and execution time almost 18 minutes.

As you see, all eight nodes are accessible in this moment. The Table shows us, in A2 column, Node8 is the best and Node1 is worst (in fuzzy range). When these results have been delivered to server, broker provider on server side has selected Node4 for

Table 3. The computed result in 7.00 to 9.00 o'clock

Node's Name	A1	A2	A3	A4	A5	A6	A7	A8	A9
Node$_1$	0.89	0.9	0.9	0.9	1	0.55	0	0.54	0.02
Node$_2$	0.87	0.94	0.9	0.85	0.95	0.8	0.21	0.77	0.04
Node$_3$	0.9	0.92	0.9	0.85	1	0.67	0.39	0.97	0.05
Node$_4$	0.94	0.95	0.95	1	1	0.9	0.48	0.87	0.43
Node$_5$	0.95	0.96	1	0.9	1	0.85	0.48	0.98	0.39
Node$_6$	0.96	0.95	1	1	1	0.7	0.48	0.16	0.38
Node$_7$	0.92	0.93	0.9	0.95	1	0.85	0.48	0.8	0.28
Node$_8$	0.8	1	0.85	1	1	1	0.78	0.65	1

this purpose. Then job sent to this node for execution and after a little time, job finished successfully on Node4.

The priority list nodes for this job were as following (high reliability priority for job): $Node_4 > Node_5 > Node_8 > Node_7 > Node_6 > Node_3 > Node_2 > Node_1$

If this job had a real-time priority, the below order was selected by broker provider Application: $Node_8 > Node_4 > Node_5 > Node_7 > Node_3 > Node_2 > Node_6 > Node_1$

In following days, all measures were based-on broker provider application. After doing 120 measures, we took a below results (Table 4) to execute a job with 10.24 MB and 22 minutes for approximate time. The following results sent by each participated nodes at time 12-15:

Table 4. The computed result in 12.00 to 15.00 in 21 August (job size 10.24 MB at 12:45 o'clock)

Node's Name	A1	A2	A3	A4	A5	A6	A7	A8	A9
Node$_1$	0.902	0.87	0.95	0.94	0.8	0.45	0	0.78	0.02
Node$_3$	0.908	0.93	1	0.86	0.8	0.53	0.39	0.92	0.05
Node$_5$	0.961	0.96	0.95	0.89	1	0.89	0.48	0.96	0.39
Node$_6$	0.964	0.93	0.95	0.94	0.9	0.66	0.48	0.80	0.38
Node$_7$	0.92	0.94	0.95	0.92	0.6	0.91	0.48	0.93	0.28
Node$_8$	0.81	1	0.9	0.89	0.8	1	0.78	0.75	1

As you see, there are only 6 nodes in available. This job is considered as a Real-time job, thus Node$_8$ was selected as the best node by proposed application. The selection priority of nodes is as following:

$Node_8 > Node_5 > Node_6 > Node_7 > Node_3 > Node_1$

If the job had a very high reliability priority, then the selection priority will be as following: $Node_5 > Node_8 > Node_6 > Node_7 > Node_3 > Node_1$

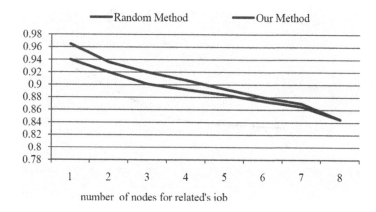

Fig. 3. The ratio of successfully finished jobs in both Random method and our provider

In this method we are choosing the best conditions for job. Whereas in other methods (for example, random methods), It's possible that have a high risk to select a node. The Ratio of successful jobs in our methods is compared with another random method [16] in below diagram (Fig 3).

In this comparison for each node, we have considered one job and the results of testing have been repeated for many times. As you see, there are eight nodes and if one node is selected, this best condition in our approach and with increasing the number of selecting nodes, our method will be similar to Random method. So, this diagram shows that the selected nodes by our method have good performance in stable state.

Finally, the last diagram (Fig 4) illustrates the time execution for several jobs with different size over 4 nodes. At the first time the proposed applications is executed to selecting best nodes. This is maybe take time less than 1 minute to a few minutes (depend on nodes positions and network bandwidth).

As you see, when the job size and time execution is increased, the random method has not operated as well as job that has small size an small estimation time. Because selected nodes in random method is different from our method. On the other hand, maybe this node for small jobs (small size and execution time) had been best efficiency. Thus, it is possible that random method select this node for a desired job.

The result shows that our approach can achieve better performance under this strategy. After each measure, it is seemed that, the ratio of successfully finished jobs, have improved. It is memorable that, for all jobs smaller than 5MB and approximate time less than 2 minutes, almost all jobs finished successfully. Therefore if there is a suitable method according to job requirement for selecting optimal nodes, performance of resources will increase because we can select some nodes that they were best nodes (in efficiency and availability) in the past. Also, time execution (for heavy processes) and also fault rate will be decreased, because we can select some nodes that they have much Free RAM and CPU-Idle with few processes.

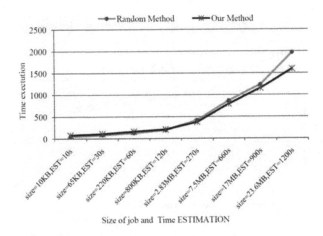

Fig. 4. The ratio of successfully finished jobs in both Random method and our provider

6 Conclusion and Future Works

Dynamic and Instantaneous Selecting some nodes for scheduling in the pool of discovered nodes is a challenging problem. Many methods for this purpose have been presented but all of them have some restrictions. We have shown that our method is a viable contender for use in future Grid implementations. This is supported by the experimental results obtained on a Grid. Our proposed approach is learning based with high accuracy which can reduced extra overhead communications and faults in cycle of selection. This broker provider application along with MA offer a dynamic decision to access any of the available and appropriate nodes by using main important criteria. Considering this characteristic we recommend this method is useful in cases in which we have sufficient nodes. For the future work we decide use a dynamic framework for the broker on economic-based grid. In this way, depend on staus of jobs and grid, The broker can select a specific nodes with considration of cost, time execution,expiration time, and so on.

Acknowledgments. This work was funded by Islamic Azad University-Miyandoab branch. The authors would like to thanks to Mr. Ali Imani and Mr. Hosein Rajabalipoure at University Technology of Malaysia (UTM), for their helps in preparation of this paper.

References

1. Czajkowski, K., Fitzgerald, S., Foster, I., Kesselman, C.: Grid information services for distributed resource sharing. In: 10th IEEE Symposium on High Performance Distributed Computing, San Francisco, California, August 7-9 (2001)
2. Krauter, K., Buyya, R., Maheswaran, M.: A Taxonomy and Survey of Grid resource Management Systems. Software Practice and Experience 32(2), 135–164 (2002)

3. Umano, M., Okamoto, H., Tamura, H., Kawachi, F., Umedzu, S., Kinoshita, J.: Fuzzy decision trees by fuzzy id3 algorithm and its application to diagnosis systems. Department of Systems Engineering and Precision Engineering, Osaka University, Japan, IEEE (1994)

4. Kim, Y., Yu, J., Hahm, J., Kim, J., et al.: Design and Implementation of an OGSI Compliant Grid Broker Service. In: Proc. of CCGrid (2004)

5. Dumitrescu, C., Foster, I.: GRUBER: A Grid Resource SLA Broker. In: Cunha, J.C., Medeiros, P.D. (eds.) Euro-Par 2005. LNCS, vol. 3648, pp. 465–474. Springer, Heidelberg (2005)

6. Rodero, I., Corbalán, J., Badia, R.M., Labarta, J.: eNANOS Grid Resource Broker. In: Sloot, P.M.A., Hoekstra, A.G., Priol, T., Reinefeld, A., Bubak, M. (eds.) EGC 2005. LNCS, vol. 3470, pp. 111–121. Springer, Heidelberg (2005)

7. Casanova, H., Obertelli, G., Berman, F., Wolski, R.: The AppLeS parameter sweep template: user-level middleware for the grid. In: Proceedings of the 2000 ACM/IEEE Conference on Supercomputing (2000)

8. Chapman, B., Sundaram, B., Thyagaraja, K.: EZ-Grid: Integrated Resource Brokerage Services for Computational Grids (2001), http://www.cs.uh.edu/ezgrid/

9. Klir, G.J., Yuan, B.: Fuzzy Sets and Fuzzy Logic: Theory and Applications. Prentice Hall, Englewood Cliffs (1995)

10. Marsala, C., B-Meunier, B.: Choice of a method for the construction of fuzzy decision trees. In: The IEEE International Conference on Fuzzy Systems, University P. e M. Curie, Paris (2003)

11. Janikow, C.Z.: Fuzzy decision trees: Issues and methods. IEEE Transactions on Systems, Man and Cybernetics 28 (1998)

12. Marsala, C., B-Meunier, B.: Choice of a method for the construction of fuzzy decision trees. In: The IEEE International Conference on Fuzzy Systems, University P. e M. Curie, Paris (2003)

13. Zadeh, L.A.: Making computer think like people. IEEE spectrum 8, 26–32 (1984); [8] S.Haack, Do we need fuzzy logic? Int .Jr nl .of Man-Mach.stud 11 (1979)

14. Bouyer, A., movaghar, A., arasteh, B.: A Multi Versioning Scheduling Algorithm for Grid System Based on Hierarchical Architecture. In: Proceedings of the 7th IADIS International Conference on WWW/Internet, Vila Real, Portugal (October 2007)

15. Frey, J., Tannenbaum, T., Foster, I., Livny, M., Tuecke, S.: Condor-G: A Computation Management Agent for Multi-Institutional Grids. In: Proceedings of the 10th IEEE Symposium on High Performance Distributed Computing (HPDC10), San Francisco, CA (August 2001)

16. Meybodi, M., Ariabarzan, N.: A dynamic methods for searching and selecting nodes in peer to peer fashion. In: Proc in IKT 2006: 10th conference in computer science, Tehran (2006)

17. Dumitrescu, C.: Problems for Resource Brokering in Large and Dynamic Grid Environments. In: Nagel, W.E., Walter, W.V., Lehner, W. (eds.) Euro-Par 2006. LNCS, vol. 4128, pp. 448–458. Springer, Heidelberg (2006)

18. Buyya, R., Abramson, D., Giddy, J.: Nimrod/G: An architecture of a resource management and scheduling system in a global computational grid. In: Proceedings of HPC Asia 2000, Beijing, China, May 14-17 (2000)

19. Zhao, M.-H., Chen, Y.-Z., Liu, D.-R., Li, J.: The application of optimized fuzzy decision trees in business intelligence. In: Proceedings of the Fifth International Conference on Machine Learning and Cybernetics, Dalian, China (August 2006)

A New Multidimensional Point Access Method for Efficient Sequential Processing of Region Queries

Ju-Won Song, Sung-Hyun Shin, and Sang-Wook Kim

College of Information & Communications, Hanyang University, Korea

Abstract. The B^+-tree was proposed to support sequential processing in the B-tree. To the extent of authors' knowledge, however, there have been no studies supporting sequential processing in multidimensional point access methods(PAMs). To do this, the cells in a multilevel and multidimensional space managed by a multidimensional PAM must be linearly ordered systematically. In this paper, we discuss an approach that linearly orders cells in a multilevel and multidimensional space and propose a novel sequential processing algorithm for region queries using this approach. We then propose the MLGF-Plus applying this approach to the MLGF.

1   Introduction

After the B-tree was invented by multiple teams simultaneously, the B^+-tree was proposed for supporting sequential processing of range queries without tree traversal [1]. The B^+-tree not only has all of advantages of the B-tree but also supports efficient sequential processing of range queries in one dimensional space without costly tree traversal.

In the B^+-tree, the keys of all objects are stored in leaf nodes, and all the leaf nodes are linked in increasing order of the keys. These all linked leaf nodes are called the *sequence set* [1]. The SSS(*subset of the sequence set*) for a query is defined as a linked portion of the sequence set that starts from a leaf node containing the starting point of the query range and ends a node containing the ending point of the range. A range query in one dimensional space can be processed by scanning of an SSS for the query.

In the last two decades, various multidimensional point access methods(PAMs) were studied for handling of objects that can be represented as points in a multidimensional space. As results, robust PAMs that efficiently handle various distributions including skewed, such as KDB-tree [6], MLGF [10,11,13,7,8], and LSD-tree [2] were developed. These PAMs can be regarded as extensions of the B-tree or the B^+-tree for the multidimensional space in some aspects.

Also, a method using the space filling curve(SFC) for handling multidimensional objects without the extension of the B-tree was proposed [4]. The SFC is used for ordering of grid patterned square cells of a multidimensional space.

T.-h. Kim et al. (Eds.): FGCN 2008 Workshops and Symposia, CCIS 28, pp. 206–218, 2009.

For processing of region queries, costly tree traversals are needed in a multidimensional PAM as in the B-tree. If sequential processing of region queries is supported as in the B$^+$-tree, we can expect great performance enhancements for region query processing. But, to the extent of the authors' knowledge, there have been no methods supporting sequential processing of region queries in multidimensional PAMs. In this paper, we propose an approach that extends multidimensional PAMs for supporting sequential processing of region queries. Using the approach, we can process a region query with efficient sequential accessing of multiple SSSs. An SSS is composed of linked disk blocks. The method does not lose all of current benefits of multidimensional PAMs.

This paper is organized as follows. The characteristics of SFC are explained in Section 2. In Section 3, we define extended grid pattern and SFC for handling of a multilevel and multidimensional space. Then, we show the splitting strategy of a multidimensional PAM must reflect the characteristics of the extended SFC. Next, we propose a sequential processing algorithm of region queries in a multidimensional space using the extended SFC. In Section 4, we propose a new index structure the MLGF-Plus applying our approach of Section 3 to the MLGF. In Section 5, we conclude the results and give further research directions of the proposed ideas.

2   Characteristics of Space Filling Curves

The SFC orders the square shaped cells with full grid patterns of n dimensional space that divide each axis with the number of 2^k ($k = 1, 2, ...$) with equal length. The total number of cells is 2^{kn}. The SFC can be used for mapping of cells in multidimensional space to linear order. Typical such SFCs are the Z-order and the Hilbert-order [4]. Figures 1 and 2 show how these two SFCs order cells in the two dimensional space when $k = 1, 2$, and 3.

Generally, the query region for a region query is determined as some prefixed range or a value for some of multiple attributes. We can regard that the entire range is selected for undetermined attributes. So, a query region is a hyperrectangle. The cells intersected with the query region may be divided as m SSSs. In this case, the query may be processed m scans of SSSs.

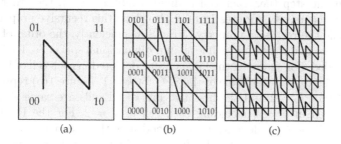

Fig. 1. The Z-order with $k = 1, 2$, and 3

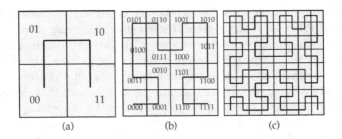

Fig. 2. The Hilbert-order with $k = 1$, 2, and 3

Since the number of scans must be reduced to process queries efficiently, the SFC must have characteristics of making a small number of SSSs for various forms of a query region. This means that the SFC must have a property of **degree of locality** as high as possible. So, the SFC must locate cells in the linear order as adjacently as possible for adjacent cells in the multidimensional space to reduce the number of scans for query processing. This is also because that the processing may be handled with one scan if the distance between two scans is not too far.

The reasoning behind the locality is that the cells must be located in the linear order as adjacently as possible for adjacent cells in the multidimensional space. For example, for the Hilbert-order of two-dimensional space, two cells among four adjacent cells are located at contiguous positions in the linear order and the average and variance of the distant value for the discontiguous two cells are relatively small. But, in the Z-order, only a cell of four adjacent cells is contiguous in some cases and the average and variance are relatively large. In [4], it is discussed that the Hilbert-order has the best property with theoretical and experimental investigation among the SFCs.

In the tree structured multidimensional PAMs, a region corresponding to a parent node of the tree includes all regions corresponding to all leaf nodes pointed by the parent node. If an SFC is applied to a multidimensional space managed by a parent level of the tree, **the SFC used for the child level must be generated from the SFC used for the parent level recursively.** In other words, the SFCs must have a property of the fractal. If the granule of cells is reduced in one step (i.e., each cell is divided into 2^n cells), the order must be maintained recursively. Figures 1 and 2 also show this recursive property.

To handle the recursive property of the SFC efficiently, the order of each cell can be represented n bits for each level in n dimensional space. So, for l levels, nl bits are needed. For example, the binary value depicted in each cell of Figure 1(a) represents the cells' order value of the SFC for $k = 1$. Figure 1(b) represents the order value for $k = 2$ and the first two bits of all cells are same as the order value of a cell in $k = 1$ including each cell in $k = 2$. For the Hilbert-order, the characteristic is analogous. This method uses the lexicographical order for representing the order value of cells. In this method, the inclusion and adjacency relation among cells can be easily determined.

3 Multifarious Granuled Grid(MGG) and Multifarious Granuled SFC(MGSFC)

In Section 3.1, a new grid pattern usually used for managing the multidimensional space in multidimensional PAMs is defined. In Section 3.2, we define an extended SFC that can be applied to the new grid pattern and discuss that the splitting strategy used for PAMs must reflect the characteristics of this extended SFC. In Section 3.3, we propose a sequential processing algorithm using recursion for region query in multilevel and multidimensional space.

3.1 Multifarious Granuled Grid(MGG)

In Section 2, we only consider the full grid pattern that is constructed by the same sized square cells. The grid file [3] is an index structure managing objects in multidimensional space with a full grid pattern. If many cells point to the same disk block, then the cells may be unified to one cell. When multidimensional space is managed by a full grid pattern, the number of cells becomes too large for managing data with skewed distribution [10]. So, it is preferable to define the extended form of a grid pattern managing skewed distribution efficiently.

To reduce the complexity of managing the multidimensional space, we assume that the bounding lines of a cell are parallel to the axes, the shape of all cells is rectangle, and the start and ending position of all cells is $p/2^n$ ($p=0,1,..,$ $2^n - 1$) and $q/2^n$ ($q=1,2,.., 2^n$), respectively: i.e., the shape of all cells is the hyper-rectangle. We define this type of a grid pattern as *Multifarious Granuled Grid(MGG)*. A representative multidimensional PAM handling the MGG is the MLGF [7,10,13].

Although cells in MGG may be splitted to be various styles, we only use a style. If the overflow for the disk block pointed by a cell occurs, we select an axis of the cell for splitting and divide the center of the range for the axis with a restriction of selecting the axis that is not selected yet from the square shaped cell for always maintaining square like shaped cells. This means that the length of an axis for a cell is equal or twice to the length of the other axes for the cell.

3.2 Multifarious Granuled Space Filling Curve(MGSFC)

As explained in Section 2, the SFC must be extended to order the cells of the MGG since the MGG does not have the grid pattern. Also, the extended SFC must have the properties explained in Section 2.

Since the SFC has the recursive property, the SFCs in multilevel space may be integrated according to the granules of the cells. So, we can order the cells of the MGG with the extended SFC. We call the extended SFC as *Multifarious Granuled Space Filling Curve(MGSFC)*. We also use the same method of representing the order value of the cells as explained in Section 2[1].

---

[1] This method can be used in multidimensional space. The outline of extending the Hilbert order for the three dimensional space is explained in [4].

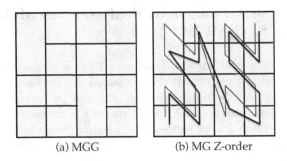

(a) MGG   (b) MG Z-order

Fig. 3. An example of the MGG and the MG Z-order

The MGSFC and the splitting strategy are very closely related. For example, Figure 3(a) shows an example of the MGG that uses the round-robin splitting strategy in two-dimensional space. The thin line in Figure 3(b) shows the Z-order of a grid pattern with $k = 2$, and the bold line shows the MG Z-order applied for the MGG of Figure 3(a). As we can see, the MG Z-order can be regarded as an integration of the Z-orders of multilevels. So, the MG Z-order is closely related with the round-robin splitting strategy.

In the MGG, the cells of a non-square shape can be divided into multiple sub-cells of a square shape and the order values of these sub-cells in the SFC must be consecutive. For example, the cell in the upper left corner in Figure 3(a) can be divided into two sub-cells of a square shape and the two sub-cells are consecutive in the Z-order with $k = 2$. In such a case, we can regard the consecutive order values in the SFC as one order value in the MGSFC. For example, the order value of the cell in the upper left corner in Figure 3(a) can be represented as '010' that includes the prefix of the order values of the two sub-cells '0100' and '0101'.

The relation of the inclusion and adjacency of cells in the MGG can be easily determined using the method of representing the order values of the cells explained in Section 2. For the two cells with order values of the different length, the shorter cell is included in the longer cell if the prefix bits of longer cell is same as all the bits of the shorter cell. For example, the cells with order values '0100' and '0101' are included in the cell with an order value '010'. When the order value of the longer cell is increased or decreased by one, the two cells are adjacent if the prefix bits of the longer cell is same as all the bits of the shorter cell. For example, the cells '0011' and '01' are adjacent since the cells '0011' and '0100' are adjacent and the cell '0100' is included in the cell '01'. Also the cells '0100' and '00' are adjacent since the cells '0100' and '0011' are adjacent and the cell '0100' is included in the cell '01'.

The splitting strategy for the MGG must reflect the characteristics of the MGSFC. To use some MGSFC, we must use an adequate splitting strategy. Figure 4(a) shows an example of one to one correspondence with the MGG and the MG Hilbert-order.

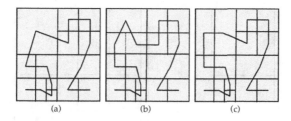

Fig. 4. One to one correspondence between the splitting strategy and the MGSFC: (a) MGG and MG Hilbert-order, (b) improperly split case, (c) properly split case

To use some MGSFC, a proper axis must be selected as a splitting axis when a cell of the MGG is split. For example, we can't apply the MG Hilbert-order when the upper left cell of Figure 4(a) is splitting as the Figure 4(b). This is because the divided lower cell has two sub-cells (the dotted line divides the cell) with not adjacent values '0100' and '0111'. So, the cell can't have one order value in the MG Hilbert-order. As explained above, the cell in the MGG must include conservative ordered sub-cells in the SFC. On the other hand, the two upper and lower divided cells include two sub-cells with adjacent order values when the upper left cell of Figure 4(a) is split as in Figure 4(c). So, this split is adequate.

We must use the splitting strategy that has a property that the sub-cells of splitted cells must be adjacent in the SFC. We discussed it with the example of Figure 4. This principle also can be applied to the case of Figure 3 using the round robin splitting strategy and the MG Z-order.

Therefore, the splitting strategy must reflect the property of the adapted MGSFC. So, we can linearly order all the cells of multilevel and multidimensional space using the appropriate MGSFC and splitting strategy. Using the observations, we can process region queries sequentially.

3.3   Region Query Processing

We only consider the case that the multidimensional space is managed by multi-levels and each level of multidimensional space can be represented by the MGG when the space is handled by a multidimensional PAM. In this case, the PAM should have a balanced tree structure and the database for an upper level is an abstract database of its lower level: i.e., a cell in an upper level may be divided into multiple cells in a lower level. It has one to n relationship. For example, in Figure 5 which represents MGGs of two levels, the cell 'x' of an upper level at Figure 5(a) is divided into cells 'A', 'B', and 'C' of its lower level at Figure 5(b).

When a query region is given for a multidimensional PAM, for each level, all pages corresponding to the cells intersected with the query region are accessed for query processing. For example, the pages corresponding to cells 'x' and 'z' are accessed for the query region 'Q' for the upper level. The sum of these cells is

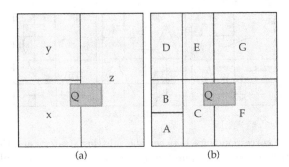

Fig. 5. The MGG of upper and lower levels

larger than or equal to the query region. We call the sum as *MGG corresponding region* of a query region.

However, the pages corresponding to the lower level cells that are divided from all the upper level cells within the MGG corresponding region for a query region may not be accessed for query processing. This is because that some cells in a lower level divided from its upper level cells within the MGG corresponding region may not be intersected with the query region: i.e., the MGG corresponding region of the upper level is larger than or equal to the region of its lower level. For example, in Figure 5, the page corresponding to the cell 'x' must be accessed for query processing but the page corresponding to the only cell 'C' (not all the cells 'A', 'B', and 'C' divided from the cell 'x' of the upper level) is accessed for query processing.

An MGG corresponding region for each level may be divided into multiple regions corresponding to multiple SSSs. This is because that the cells corresponding to MGG corresponding region may be divided into regions corresponding to multiple SSSs.

An SSS of the upper level may be divided into multiple SSSs of the lower level. The MGG corresponding region of the lower level may be smaller than that of the upper level. So, the cells corresponding to the shrinked region of the lower level for query processing may not be adjacent in the MGSFC order of the lower level since the region of the lower level corresponding to the given upper level SSS may be smaller than the region of the upper level corresponding to a given SSS.

With above observations, Algorithms 1-3 show the pseudo code of our recursive algorithm of handling region queries sequentially for the multilevel MGG environment. In the algorithms, we assume that the cursor point of the current handled entry of each level is managed with global variable and all entries for each level are linked as a chain style according to the MGSFC order value. We also assume that the page corresponding to a cell in the lowest level points the leaf page saving key of objects.

In this algorithm, an SSS is constructed in the root level at first and a lower level SSS is constructed. This process is done recursively. Then, leaf level pages

---

Algorithm 1. process_region_query
___
Input: Q
1: $level = 0$ // root level
2: Set the cursor of an entry chain of the root level as the first entry of the root level entry chain.
3: **while** (1) **do**
4:    **if** generate_next_sub_sequenceset(Q, *level*, *start_entry*, *end_entry*) returns null **then**
5:       break
6:    **else**
7:       process_ below_subsequenceset(Q, *level* + 1, *start_entry*, *end_entry*)
8:    **end if**
9: **end while**

Algorithm 2. generate_next_sub_sequenceset
___
Input: Q, *level*
Output: *start_entry*, *end_entry*
1: Find the first entry that intersects with Q from the current entry cursor position of the level to the next entries of the entry chain, and set the entry as *start_entry*.
2: **if** there exists no such entry in the entry chain **then**
3:    return null
4: **end if**
5: Find the entry that does not intersect with Q and set *last_entry* as the entry previous to this entry, and set the current entry cursor position of *level* as the entry next to this entry.
6: return

Algorithm 3. process_below_subsequenceset
___
Input: Q, *level*, *s_entry*, *e_entry*
1: **if** the level is a leaf directory *level* + 1 **then** // data page level
2:    Read the data page pointed by *s_entry* and check the objects in the page whether they satisfy the query and do the same thing for the data pages pointed by the chain to the *e_entry*
3: **end if**
4: Read the directory page pointed by *s_entry*.
5: Set the first entry of this directory page as the current cursor position of *level*.
6: **while** (1) **do**
7:    Find the first entry that intersects with Q from the current entry cursor position of level and following the entry chain, and set the entry as *start_entry*.
8:    **if** there exists no such an entry in the entry chain **then**
9:       return
10:    **end if**
11:    Find the entry that does not intersect with Q and set *last_entry* as the entry previous to this entry, and set the current entry cursor position as the entry next to this entry.
12:    process_below_subsequence_set(Q, *level* + 1, *start_entry*, *end_entry*)
13: **end while**

---

having key values of objects are accessed for the lowest level. This algorithm adapts the depth-first processing techniques[2]. So, for each level, only an SSS is handled for query processing. The breath-first technique may be used. The depth-first technique would be more efficient.

4   MLGF-Plus

The approach proposed in Section 3 is a general method that can be applied for all PAMs managing the MGG. But, for applying the approach to a certain PAM managing the MGG, the detail implementation method considering the characteristics of the PAM must be designed. In this section, we address this for applying our approach to the MLGF [7,8,10,11,13]. In Section 4.1, we explain the characteristics of the MLGF [7,8,10,11,13] that can manage the MGG. In Section 4.2, we propose a new index structure applying the approach proposed in Section 3. We call it *the MLGF-Plus*. So, the MLGF-Plus can support efficient sequential processing of region queries.

4.1   MLGF

The MLGF is a balanced tree and consists of a multilevel directory and data pages [10,11]. Each directory level reflects the status of space partition of a level. A directory page has directory entries contained in a region that corresponds to a higher-level directory entry pointing to the page. When the MLGF is used as an index, each directory entry in a leaf level consists of a key of the corresponding object and a pointer to that object. So, the MLGF is an n-ary balanced index structure basically.

A directory entry in a nonleaf level consists of a *region vector* and a *pointer* to a lower-level page. A region vector in an n-dimensional MLGF consists of n hash values that uniquely identify the region. The position, shape, and size of a region are reflected in the region vector. The i-th hash value of a region vector is the common prefix of the hash values for the i-th attribute of all the possible objects that belong to the region. Figure 6 represents the MLGF structure managing the MGG of Figure 5. For example, the directory entry E in this figure represents the region containing all the objects having common prefix '01' and '1' for the first and second attributes respectively when the two attributes of the objects are hashed.

By splitting and merging pages, the MLGF adapts to dynamic environments where insertions and deletions of objects frequently occur. If a directory page overflows, the corresponding region is split into two equal-sized regions (say, *the halving strategy*). The MLGF employs the local splitting strategy [5] that splits a region locally, rather than across the entire hyperplane, when the corresponding

---

[2] In each level, if the number of entries between two SSSs is lower than some predetermined value, we may handle these two SSSs as one SSS. With this method, the false-drop is larger but the number of scans gets smaller. We left it as a further research.

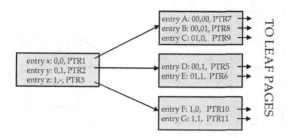

Fig. 6. Two level directory structure of two dimensional MLGF

page overflows. The local splitting strategy maintains the policy of having one directory entry correspond to one page; thus, the strategy prevents the MLGF from creating unnecessary directory entries. As a result, the directory size of the MLGF is linearly dependent on the number of inserted objects regardless of data distribution or correlation among the attributes [5]. So, the MLGF can manage the MGG.

Basically, in the MLGF, the axis for splitting is not determined when the splitting is needed. So, as discussed in Section 3, the MLGF can be generated by using the characteristics of the adapted MGSFC. Thus, the MLGF is a PAM that can manage the MGG, which proposed in Section 3.1, and can employ the query processing method proposed in Section 3.3.

4.2 MLGF-Plus

In this section, we discuss how the approach of Section 3 is applied to the MLGF. We also discuss the difference between algorithms of the MLGF and the MLGF-Plus for the region query processing.

For linear ordering of the cells of multilevel and multidimensional space that are managed by the MLGF-Plus, we maintain the order of cells by the MGSFC, *the MGSFC order value*, for each level. For doing this, we should know the MGSFC order value for each region that corresponds each entry. The order value can be stored in an additional field for each entry or can be calculated by using the region vector of each directory entry. The former is not preferable since it decreases the fan-out because of growing of the entry size. Thus, we use the latter although it needs some additional calculations.

By using these order values, we can store the entries for all regions of a directory page according to the order of the MGSFC. This can be done by using the insertion sort technique and thus requires position changes of directory entries. Although an additional field may be used without the position changes, we do not use this technique since it incurs growing of the entry size.

When the two regions corresponding to two entries K and L stored in other directory pages are adjacent in the order of the MGSFC, we can maintain the adjacency relation by storing the page ID of L in the page containing K (with

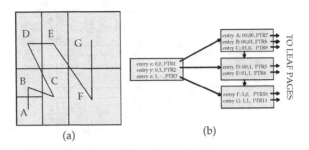

Fig. 7. An example of the MLGF-Plus

an additional field) and by some means of knowing the position of L in the page containing L. For the former, we use a new additional filed, *next neighbor page ID field*. For the latter, since the entry L has an entry having the smallest MGSFC order value in the page containing the entry L, the entry L is the first entry of the page. So, no additional field is needed. This information must be maintained always with dynamic changes of the MLGF-Plus. Figure 7(a) represents the MG Z-order for the Figure 5(b) and Figure 7(b) shows its MLGF-Plus structure that extends the original MLGF structure.

Our proposed algorithm of region query processing for the MLGF-Plus recursively processes with all the entries in an SSS for each level, while the algorithm for the MLGF [7,8,10,11,13] recursively processes with an entry for each level. The former considers the adjacency relation of the entries, but the latter does not. So the former has advantage of reducing the access number of directory pages accessed for middle levels since all directory pages for the lowest level pointed by an SSS can be processed at once. This is the advantage of the B^+-tree over the B-tree. But, our algorithm handles multiple SSSs for a given query region rather than handling an SSS as the B^+-tree.

If directory pages for an SSS are stored in a sector or adjacent sectors of the disk, we can reduce the moving time of the disk head. Also, if all leaf pages pointed by entries of a directory page corresponding to an SSS have similar conditions, we have same effects.

If data pages storing objects pointed by the leaf pages that are pointed by an SSS(i.e., multiple leaf directory pages) is physically adjacent(i.e., it means the leaf pages are clustered by the MGSFC order value), we can have greater advantages[3]. If objects are clustered with the MGSFC order value, we can process region queries with similar time for any rectangle style of a query region since the

---

[3] This method may be considered as an extension of the bulk loading method for the MLGF [12] with using the Z-order in static environments. We can enhance the performance of region query processing by using the MGSFC rather than considering the Z-order. The fixed positioned object storing system, usually used in most DBMSs, uses the concept of *near object* [8,9]. It tries to make the adjacent objects stored in the same page in disk. If the MGSFC order values are used as a parameter of adjacency, the performance of query processing can be enhanced.

MGSFC order value considers all the attributes. This method uses the MGSFC order value as the clustering attribute.

5   Conclusions

The very famous one dimensional index structure B-tree was extended to the B^+-tree for supporting sequential processing [1]. But, there are no studies for supporting of sequential processing in multidimensional PAMs.

To support query processing for region queries sequentially in multidimensional space, the cells in multilevel and multidimensional space must be ordered linearly. To do this, we can use the SFC with some extension. We explain that the SFC has properties of locality and recursion basically. There are the Z-order and Hilbert order as the very famous SFC. These orders assume the full grid pattern for ordering of cells. But the full grid pattern yields a very large index structure.

To overcome the disadvantage of the full grid pattern, there are many studies of tree structured multidimensional PAMs. These PAMs integrate cells pointing to same disk blocks as one cell [2,10]. We restrict the case that the PAMs use the half splitting strategy and have the square like cells having the length of an axis with equal or twice of the other axes. We define this type of a grid pattern as the MGG. We also discussed that an upper level MGG is an abstraction of a lower level MGG. We define the MGSFC, an extension of the SFC, for linear ordering of cells for the some case of the MGG. We also discussed that the splitting strategy used for a multidimensional PAM must reflects the characteristics of the MGSFC. This method is used for ordering of cells in a multilevel and multidimensional space to support sequential processing of region query. We present sequential query processing algorithms using this approach. The approach discussed here can be applied not only to the MLGF but also to all PAMs managing multilevel and multidimensional space with the MGG style.

We also proposed the detailed method for applying this approach to the MLGF. We call the new structure the MLGF-Plus. After implementing this approach, we will study the advantages and disadvantage of our approach via extensive experiments.

Acknowledgements

This work was supported by the Korea Research Foundation Grant funded by the Korean Government (MOEHRD, Basic Research Promotion Fund) (KRF-2007-314-D00221), and was partly supported by the MKE(Ministry of Knowledge Economy), Korea, under the ITRC(Information Technology Research Center) support program supervised by the IITA(Institute of Information Technology Advancement) (IITA-2008-C1090-0801-0040).

References

1. Comer, D.: The Ubiquitous B-tree. Computing Surveys 11(2), 121–137 (1979)
2. Henrich, A., Six, H.W., Widmayer, P.: The LSD Tree: Spatial Access to Multidimensional Point and Nonpoint Objects. In: Proc. 15th Intl. Conf. on Very Large Data Bases, pp. 45–53 (1989)
3. Hinrichs, K., Nievergelt, J.: The Grid File: A Data Structure Designed to Support Proximity Queries on Spatial Objects. In: Proc. Intl. Workshop on Graph Theoretic Concepts in Computer Science, pp. 100–113 (1983)
4. Jagadish, H.J.: Linear Clustering of Objects with Multiple Attributes. In: Proc. Intl. Conf. on Management of Data, ACM SIGMOD, pp. 332–342 (1990)
5. Kim, S.W., Whang, K.Y.: Asymptotic Directory Growth of the Multilevel Grid File. In: Proc. Intl. Symp. on Next Generation Database Systems and Their Applications, pp. 257–264 (1993)
6. Robinson, J.T.: The K-D-B Tree: A Search Structure for Large Multidimensional Dynamic Indexes. In: Proc. Intl. Conf. on Management of Data, ACM SIGMOD, pp. 10–18 (1981)
7. Song, J.W., Whang, K.Y., Lee, Y.G., Lee, M.J., Kim, S.W.: Spatial Join Processing Using Corner Transformation. IEEE Trans. on Knowledge and Data Engineering (IEEE TKDE) 11(4), 688–695 (1999)
8. Song, J.W., Whang, K.Y., Lee, Y.G., Kim, S.W.: The Clustering Property of Corner Transformation in Spatial Database Applications. In: Proc. 23rd Intl. Computer Software and Applications Conf. (COMPSAC 1999) (October 1999)
9. Song, J.W., Whang, K.Y., Lee, Y.G., Lee, M.J., Han, W.S., Park, B.-K.: The Clustering Property of Corner Transformation for Spatial Database Applications. Information and Software Technology 44(7), 419–429 (2002)
10. Whang, K.Y., Krishnamurthy, R.: Multilevel Grid Files. IBM Research Report RC 11516(51719) (1985)
11. Whang, K.Y., Krishnamurthy, R.: The Multilevel Grid File - a Dynamic Hierarchical Multidimensional File Structure. In: Proc. 2nd Intl. Conf. on Database Systems for Advanced Applications, pp. 449–459 (1991)
12. Won, H.S., Kim, S.W., Song, J.W.: An Efficient Algorithm for Bulk-Loading of the Multi-Level Grid File. In: Proc. 6th Intl. Conf. on Computer Science and Informatics (CS&I 2002), Durham, NC, USA, May 2002, pp. 382–386 (2002)
13. Whang, K.Y., Kim, S.W., Wiederhold, G.: Dynamic Maintenance of Data Distribution for Selectivity Estimation. The VLDB Journal 3(1), 29–51 (1994)

Automatic Extraction of HLA-Disease Interaction Information from Biomedical Literature

JeongMin Chae[1], JiEun Chae[2], Taemin Lee[1], YoungHee Jung[1],
HeungBum Oh[3], and SoonYoung Jung[1]

[1] Department of Computer Science Education, Korea University, Korea
bluesky@comedu.korea.ac.kr
[2] Department of Computer and Information Science, University of Pennsylvania,
USA
[3] Department of Laboratory Medicine, Asan Medical Center and University of Ulsan
College of Medicine, Korea

Abstract. The HLA control a variety of function involved in immune response and influence susceptibility to over 40 diseases. It is important to find out how HLA cause the disease or modify susceptibility or course of it. In this paper, we developed an automatic HLA-disease information extraction procedure that uses biomedical publications. First, HLA and diseases are recognized in the literature using built-in regular languages and disease categories of Mesh. Second, we generated parse trees for each sentence in PubMed using collins parser. Third, we build our own information extraction algorithm. The algorithm searched parsing trees and extracted relation information from sentences. We automatically collected 10,184 sentences from 66,785 PubMed abstracts using HaDextract. The precision rate of extracted relations reported 89.6% in randomly selected 144 sentences.

Keywords: HLA, textmining, disease, interaction information.

1 Introduction

The Human Leukocyte Antigen(HLA) system is the name of the Major Histocompatibility Complex(MHC) in humans. The HLA control a variety of function involved in immune response and influence susceptibility to over 40 diseases. It is important to find out how HLA cause the disease or modify susceptibility or course of it. This is that will help in treatment of theses disease. Over the years, a number of methods have been developed for the experimental methods or the prediction of HLA binding peptides from an antigenic sequence. The experimental methods for recognition of theses HLA are both time-consuming and cost-intensive. Computation method thus, provides a cost effective way to identify these HLA. However, it is difficult to identify the HLA due to the most complicated genetic structure in human body.

T.-h. Kim et al. (Eds.): FGCN 2008 Workshops and Symposia, CCIS 28, pp. 219–230, 2009.

HLA perform an important role in human immunity and has special allelic pairs in each person. The knowledge of the alleles of HLA's main 6 genes (HLA-A, -B -C, DRB1, DQB1, DPB1) is continually developing and it was reported in 2004 that 1,729 alleles were found. However this number has been exponentially increasing every year. A person's allelic makeup can influence their response to disease. Even though a person might be infected with the same microorganism, their responses may vary from self-healing to serious disease. Because HLA allele frequency differs according to geographic location, considerable number of studies has carried out into the relationship between HLA allele frequency and disease but still little is known. Relation between HLA and IE is found though textmining technique such as Named Entity Recognition(NER) and Information Extraction(IE).

There have been various attempts to efficiently find entities within biomedical literatures. Hanisch[1] found protein names that appear in biomedical text using search terms of protein names. Hatzivassiloglou[2] and Kazama[3] used machine learning approaches with word formation pattern, POS information, semantic information, prefix, suffix, and et al. The performance of these methods is about 60-80%.

There have also been numerous attempts to find interactions between entities used in literature. Friedman[4] and Temkin[5] extracted protein-protein interactions in biomedical abstracts using keywords and grammars built by domain experts. Leroy[6] used Finite State Automata(FSA) with closed words, and demonstrated that FSA can extract information in literature. McDonald[7] generated a potential parse tree using their parser and filtered out parse trees with little information. Filtering algorithm are used to select informative parse trees with valid interaction information among potential parse trees. This method has the advantage that grammar is not necessary to extract information. Horn[8] extracted interaction information between protein and point mutations rather than extracting information between proteins. Novichkova[9]introduce a general biomedical domain-oriented system that can extract various biomedical information.

In this paper, to deal with the HLA names variants, we build the regular expression of HLA and used MeSH ontology. In this study, we intended to extract interaction information between HLA and disease using textmining methods. we make use of the structural information of the sentences with aim of finding interactions between HLA and disease. The structural information of a sentence is derived through applying parse tree to the dependency relationship of the keywords in the sentence. The systems of McDonald[7] uses the potential parse tree using their parser while our system uses the parse tree through the dependency relationships between the keywords. This method analyzes more effectively involved sentence and extracts more accuracy relation information between entities which consists of a coordinating conjunction, 'and' and 'or', etc.

Our system is divided to 5 sub-processes: Tokenizing, Pos tagging, Entity Recognizing, Syntactic Analysis and Semantic Analysis. While HaDextract system incorporated all 5 sub processes including hidden relation, other data mining systems incorporated only parts of the total process. In addition, the accuracy of HaDextract was 89.6% that is above the average accuracy of70-80% other systems.

2   Entity Recognizing in HaDextract System

In order to extract HLA-disease interaction information, HLA and disease entities in the literature need to be recognized in advance. Entities in the biomedical domain show various surface realizations, which are called term variants. To deal with the term variants, this thesis builds the regular expression of HLA and used Mesh ontology and disease abbreviation patterns.

Even though HLA and disease entities are essential components in interaction information, coexistence of HLA and disease entities within the same sentence does not always guarantee their relation. Lexicalized information in sentence also needs to be introduced. We used 25 words as 'relation keywords' and 5 words as 'filtering keywords'. Keywords such as 'Associate' and 'Correlate' showed a relation between HLA and disease, and filtering keywords such as 'Restricted' and 'specific' are used to filter sentences that do not contain HLA-disease interaction information.

Geographic locations entities was not necessary to extract interaction information, but both were used to summarize interaction information. In addition, we used abbreviations of disease entities found in literature to improve recall of disease recognition.

2.1   HLA Entity

There are three naming methods to indicate HLA entities: Antigen, Allele and Gene group. The same HLA could be displayed differently depending on its naming method. In table 1, it shows various instances named by serology, DNA, and group antigen.

Instances named by serology antigen and DNA allele show similarity in expression. For example, 'A2' in Serology antigen appears as 'A*02' expression in DNA antigen. In both naming methods, they have common that 'A2' appears after 'HLA-' keyword. We built regular expressions in table 2 to find HLA in serology antigen and DNA allele naming method.

Instances named by group antigen are different in expression with Serology and DNA Antigen since Group Antigen is the naming method that focuses in combination of alleles rather than specification of Allele. We used simple keyword matching (dictionary-based approach) to find Group antigen in literatures.

Table 1. Antigen, Allele, and Gene group

Class	Examples
Antigen	HLA-A2, A2, -A2, A2-transgenic, Bw4, Bw6, Cw1, DR1, DQ1
Allele	HLA-A*02, -A*02, A*0201, A*020101, A*02010101, A*02010102L, HLA-A*0201-transgenic
Gene group	A1CREG, A2CREG, Bw4CREG, HLA-A, A*, HLA-A*

Table 2. Regular expression for finding antigen and allele

Class	Regular Expression
Antigen	$/\backslash b((|HLA-)(A|B|Bw|Cw|DPw|DQ|DR|DRw|Dw))(\backslash d+)\backslash b/$
Allele	$/\backslash b((|HLA-)([ABCEFGHJKLNPSTUVWXYZ]|DRA|DRB\backslash d|$ $DQA[12]|DQB[123]|DO[AB]|DM[AB]|DPA[123]|DPB[12])|TAP[12]|$ $PSMB[89]|MIC[ABCDE])\backslash*(\backslash d+)([LNSCAQ]?)\backslash b/$

2.2   Disease and Geographic Locations Entity

HLA entities activate disease on some specific human type. Normally human type depend on geographic location. Therefore, Recognition geographic location entity is key factor on analysis HLA-disease interaction information.

We recognize the disease and geographic location entities in abstracts by using diseases and geographic location category of MeSH. MeSH is the ontology that provides disease and geographic location entities including 23,000 terminologies. Abbreviation and synonyms provided by MeSH enable the system to find variations in terminologies. Geographic location entity category is displayed in Fig.1.

MeSh's geographic location information about cities and the countries all over the world us to summarize HLA-disease interaction by location since the relationship between HLA and disease is different according to geographic location.

2.3   Abbreviation of Disease Entity

Variations in the abbreviation of disease become a challenge for automatic information extraction. Even the same abbreviation in publications could denote different concepts by different authors. The recognition of abbreviations has an influence on the performance of systems.

We recognize abbreviations in the literature by using abbreviation formation patterns. Most abbreviations consist of capital letters and are wrapped by parenthesis. They follow a predictable pattern, in which one letter in abbreviations

Geographic Locations (3436)

Africa (160)
Americas (908)
Antarctic Regions (0)
Arctic Regions (0)
Asia (775)
Atlantic Islands (5)
Australia (30)
Cities (108)
Europe (1011)
Historical Geographic Locations (330)
Indian Ocean Islands (3)
Oceania (46)
Oceans and Seas (5)
Pacific Islands (55)

Fig. 1. Geographic Location Entity Category

corresponds to the first letter of each word in its entirety. Schwartz[10] proposes an algorithm using these patterns to recognize abbreviations from biomedical publications. We improve upon the algorithm of Ariel and found abbreviations of diseases in abstracts.

3   Information Extraction in HaDextract System

Overall system architecture is displayed in Fig.2. 'Import Abstract Component' downloaded abstract XML file from PubMed. 'Tokenizing Component' split abstract text into words. 'POS Tagging Component' found POS of each word using fnTBL POS Tagger offered by Ratnaparkhi[11]. FnTBL was trained with GENIA Corpus 3.0 to search suitable POS in the biomedical domain. 'Entity Recognizing Component' searched entity names using regular expression and MeSH keywords, and 'Syntactic Parsing Component' created parse tree using Collins Parser[12]. 'Semantic Interpret Component' extracted HLA-disease interaction information using extracted entities, parse trees, relation keywords, and filtering keywords.

3.1   Relation and Filtering Keyword

In an attempt to find HLA-disease relation information, we developed relation and filtering keywords determined by domain experts from 309 HLA publications.

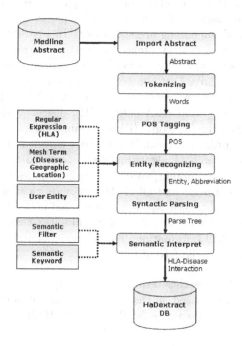

Fig. 2. HaDextract System Architecture

Table 3. Stem of filtering and relation keyword

Class	Keywords stems
Filter	bound, restricted, specific, tetramer, transgenic
Relation	associat, correlat, confer, decreas, development, differen, effect, role, found, overexpress, overrepresent, frequent, greater, higher, increas, linkage, lower, marker, predispos, progress, resist, protect, risk, suscept, effector

Domain experts found words indicating HLA-disease relation in the literature and we called those words relation keywords. However, in sentences containing HLA-disease entities and even relation keywords, it was shown that sentences describing immune responses have little HLA-disease relation. In these sentences, HLA is used for only supplementary information about genes and has no direct connection with disease. We extract words indicating immune response test and called those words filtering keyword. We use stems of keywords rather than keywords by themselves since relation and filtering keywords may appear in many different forms, as in 'Correlate', 'Correlation', and 'Correlated'. We shall discuss how to use the filtering keywords in table 3 in the information extraction section.

3.2 Parse Tree Search Algorithm

We started extracting information by generating parse trees for each sentence in PubMed using collins parser. We selected Collins parser because collins parser shows the highest precision among present parsers. In a parse tree, each terminal node represents each word in the sentence and each nonterminal node represents grammatical and dependency information between terminal nodes.

After building parse trees, we constructed our own information-extracting algorithm using postorder traversal. To extract only HLA-disease relation in a sentence, we first recognized HLA entities (H), disease entities (D), relation keywords (A), and filtering keywords (F) in terminal nodes. Then, we searched parse trees using postorder traversal. In postorder traversal, terminal nodes are searched first. When searching algorithm visits terminal nodes recognized as H, D, A, and F, it copies content(word) in terminal nodes to all of its parent nodes. If a nonterminal node had all H, D and A in its subtree, H, D and A words would be collected together in the nonterminal node. Then the searching algorithm visits the nonterminal node and it will extract the essential relation information, namely HaD information. The following enumerated steps are depicting the sample run on the tree of Fig.3.

1. Searching algorithm searches terminal nodes first.
2. Celiac disease is recognized as Disease entities and it is copied to all of its parent nodes. (celiac disease → NPB → S1 → SBAR → VP → VP → VP → S2)

3. Associated is recognized as relation keyword and it is copied to all of its parent nodes. (associated → VBN → VP → S1 → SBAR → VP → VP→ VP→ S2)

4. HLA-DQ2, DQA1*0501 and DQB1*0201 are recognized as HLA entities and they are copied to all of their parents. (HLA-DQ2, DQA1*0501 and DQB1*0201 → NPB → PP → VP → S1 → SBAR → VP → VP→ VP→ S2)

5. When searching algorithm reaches the right end of terminal nodes (DQB1*0201), it goes up to nonterminal nodes from right to left. (NPB → PP→ VP → VP → S1)

6. When searching algorithm visits S1, the first nonterminal node with HaD information, it returns the interaction information (Celiac disease associated HLA-DQ2, DQA1*0501 and DQB1*0201) and the whole procedure is terminated.

Unlike HaD words, filtering keywords play a different role in the information extracting procedure. Filtering keywords keep the searching algorithm from extracting information from little informative sentences like the immune response test. As filtering keywords are common words and have meaning only when they appear near HLA entities, the algorithm checked where filtering keywords appears in a parse tree. If filtering keywords appear closer to HLA entities than

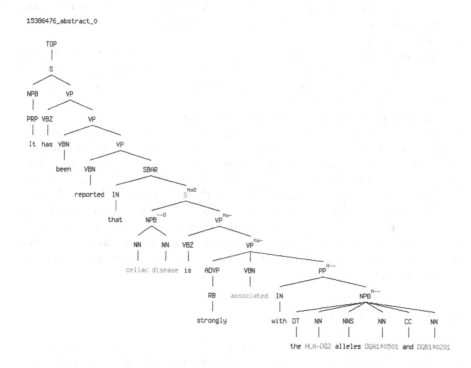

Fig. 3. Simple information extraction process

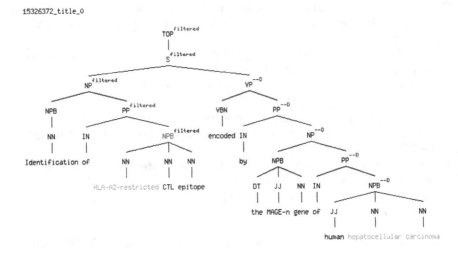

Fig. 4. Filtering process by '-restricted' as filtering keyword

other entities, the whole procedure becomes terminated without extracting any HLA-disease relation. However, if filtering keywords appear closer to disease entities (D) and relation keywords (A) than HLA entities, it loses its own filtering function, and the information is extracted even though the sentence contains filtering keywords.

Fig.4 shows filtering process by 'restricted' filtering keyword. Since filtering keywords '-restricted' appears closer with HLA entity ('HLA-A*02') than other entity ('hepatocellular carcinoma'), the sentence is filtered and searching algorithm is terminated immediately.

Compared to the method simply identifying the relation by coexistence of the HLA entities, disease entities and relation keywords, the tree search algorithm has its own advantage in that it could use structural information of sentences provided by Collins parser during filtering process. For example, the following sentence will be filtered out with simple keywords matching since filtering keywords 'specific' appear closer to HLA B27 than disease entities : 'Laboratory anomalies are not specific, HLA B27 antigen is not associated with this syndrome'. However, 'specific' and 'HLA' belong to different sentences that is connected with conjunction 'and' and in that case 'specific' do not have any dependency with 'HLA B27'. The tree search algorithm of our system did not filtered the sentence.

4   System Performance

We randomly select 909 abstracts from PubMed using search keyword 'HLA'. We carefully divided sentences in abstracts into three levels according to their quantity of information. The sentences containing HLA and disease entities at the same time are denoted as level 1. 305 sentences in 909 abstracts are denoted as level 1. The sentences in level 1 are divided again according to whether they have relation information between HLA and diseases. If sentences don't have this relation information, the sentences are denoted as level 2. If sentences have the relation information, the sentences are denoted as level 3. 144 sentences in 305 sentences (level1) were confirmed as level 3 by a domain expert.

We tested the algorithm with 305 sentences 144 sentences in level 3. Finding 129 sentences as level 3, our information extraction system reported a precision rate as 89.6%. We analyzed 15 misclassified sentences. In 4 sentences, words are incorrectly recognized as HLA entities. For example, 'A375' is misidentified as HLA entity by the system due to its similarity in expression with HLA entities. In the similar way, words are incorrectly recognized as disease entities in 2 sentences. 7 sentences without relation information did not filter out during the process.

The system showed accuracy of 57.4% in summarization. 74 sentences in 144 sentences are correctly summarized. We applied strict criterion on the evaluation of the summarization. Even if the algorithm missed any of entities or relations in a sentence, we consider it as incorrect summarization. We also analyzed the 55 inappropriately summarized sentences: we failed to find all HLA entities in 7 sentences, disease entities in 12 sentences, disease abbreviation in 4 sentences, and HLA haplotype in 11 sentences. We failed to extract information due to incorrect parsing tree in 14 sentences. Finding disease entities and disease abbreviations shows the highest error rate in our analysis. The reason of the failure is that searched disease entities were not complete due to the limitation of MeSH.

5   Results and Discussion

We collected 16,833 sentences containing HLA and disease entities at the same time from 66,785 Pubmed abstracts which contain HLA keywords from 1979 to 2004. We found 6,654 sentences are in level 2 and 10,184 sentences are in level 3. Therefore we collected 10,184 HLA-disease interaction information from 66,785 abstracts in PubMed automatically and offered it with its summary information at our web site[1].

The summary information is consisted of keywords of HLA, Action and Disease, which are found in the previous data mining process. The information is divided into three categories: HLA, disease and geographic location. Each entity in each category has HLA-disease interaction and we counted the number of relevant sentences including relevant sentences of sub entities. For example, HLA-A has A*02(1974) as sub entity. 1974 show the count of relevant sentences of HLA-A*02. HLA A*02 has A*0201(296), A*0202(1), A*0203(3), A*0205(3), A*0206(5), A*0207(12) and A*0211(2). The reason why the total count of sub entities (322) does not match with the count of A*02(1974) is that A*02 has its own count of relevant sentences. Thus, the HLA A*02's count of relevant sentences will be 1652(=1974-322). We collect 23,438 sentences on HLA, 53,875 sentences on Disease and 3,436 sentences on geographic location. Fig.5 shows

---

[1] HaDextract system (http://dataknow.korea.ac.kr/hadextract/index.php)

HLA (23438) - filtered

HLA-A
A*00(1), A*01(445), A*02(1974), A*03(354), A*04(38), A*05(38), A*06(16), A*07(13), A*08(46), A*09(104), A*10(63),
A*11(157), A*12(15), A*13(6), A*14(1), A*16(1), A*17(3), A*18(1), A*19(23), A*20(11), A*21(3), A*23(11), A*24(314),
A*25(14), A*26(52), A*27(23), A*28(46), A*29(41), A*30(27), A*31(40), A*32(7), A*33(36), A*34(2), A*37(8), A*42(3),
A*43(11), A*49(2), A*54(11), A*66(2), A*67(1), A*68(21), A*69(1), A*74(1), A*81(1)

HLA-B
B*00(1), B*01(93), B*02(27), B*03(19), B*04(99), B*05(153), B*06(31), B*07(681), B*08(988), B*09(2), B*10(57), B*11
(2), B*12(164), B*13(113), B*14(82), B*15(138), B*16(66), B*17(95), B*18(133), B*19(55), B*21(37), B*22(47), B*25(3),
B*26(1), B*27(3663), B*30(2), B*35(338), B*37(33), B*38(44), B*39(57), B*40(107), B*41(22), B*42(2), B*43(4), B*44
(119), B*45(11), B*46(61), B*47(15), B*48(6), B*49(16), B*50(4), B*51(150), B*52(84), B*53(13), B*54(53), B*55(10),
B*56(7), B*57(66), B*58(12), B*60(32), B*61(22), B*62(53), B*65(1), B*67(3), B*70(2), B*73(4), B*75(1), B*76(3), B*77
(1), B*95(21)

HLA-C
Cw*01(36), Cw*02(39), Cw*03(68), Cw*04(82), Cw*05(15), Cw*06(210), Cw*07(90), Cw*08(7), Cw*11(3), Cw*14(1),
Cw*15(8), Cw*16(6), Cw*17(1)

HLA-
DRB1
DRB1*01(604), DRB1*02(1101), DRB1*03(1860), DRB1*04(2716), DRB1*05(259), DRB1*06(80), DRB1*07(375), DRB1*08
(156), DRB1*09(139), DRB1*10(40), DRB1*11(162), DRB1*12(42), DRB1*13(168), DRB1*14(71), DRB1*15(532), DRB1*16
(48), DRB1*17(42), DRB1*35(1), DRB1*40(1), DRB1*51(8), DRB1*52(26), DRB1*53(53), DRB1*71(1), DRB1*80(1)

HLA-
DQB1
DQB1*01(50), DQB1*02(547), DQB1*03(515), DQB1*04(102), DQB1*05(157), DQB1*06(478), DQB1*07(40), DQB1*08
(201), DQB1*09(7), DQB1*11(1), DQB1*20(1), DQB1*30(1)

HLA-
DPB1
DPB1*01(17), DPB1*02(36), DPB1*03(41), DPB1*04(42), DPB1*05(30), DPB1*06(2), DPB1*09(7), DPB1*10(2), DPB1*13
(3), DPB1*14(2), DPB1*16(1), DPB1*17(6), DPB1*20(1), DPB1*21(1), DPB1*22(1), DPB1*28(1), DPB1*32(1), DPB1*35(1)

Disease (53875) - filtered

Bacterial Infections and Mycoses (1763)
Virus Diseases (2065)
Parasitic Diseases (119)
Neoplasms (5055)
Musculoskeletal Diseases (14266)
Digestive System Diseases (3275)
Stomatognathic Diseases (199)
Respiratory Tract Diseases (982)
Otorhinolaryngologic Diseases (47)
Nervous System Diseases (2755)
Eye Diseases (1371)
Urologic and Male Genital Diseases (733)
Female Genital Diseases and Pregnancy Complications (515)
Cardiovascular Diseases (1015)
Hemic and Lymphatic Diseases (1050)
Congenital, Hereditary, and Neonatal Diseases and Abnormalities (432)
Skin and Connective Tissue Diseases (5138)
Nutritional and Metabolic Diseases (1564)
Endocrine System Diseases (2027)
Immune System Diseases (6673)
Disorders of Environmental Origin (313)
Animal Diseases (16)
Pathological Conditions, Signs and Symptoms (2478)

Fig. 5. Class of extracted interaction information at whole HLA and whole disease: the number of parenthesis is the count of sentences

the total number of extracted HLA-disease interactions and categorized numbers. Extracted interaction information is displayed in Fig.6. We offer level 3 and level 2 data. Level 2 data is sometimes considered as not usable data but their statistic knowledge is important for further research. Additionally, the number of HLA-disease interactions according to geographic location demonstrates how many research studies have been carried out in those geographic locations. Furthermore, we offer some usable tools like view parsetree for all sentences, view abstracts which contain selected sentences and more.

Since HLA entity name were listed in various transformation, it is quite difficult to recognize HLA entity by using the regular expression. For example, the sentence 'Haplotype-specific patterns were defined with DNA from DR1, 2, 3, 4, 7, 8, 11, 12, and 13 homozygous typing cells, with restriction enzymes Eco RI, Bgl I, and Pvu II.' have 9 DR antigen. But we identified only DR1. This problems appeared to the sentence 'DQB1*03/*04/*05/*06, DQA1*0301 (Ls, Stage II) were less common than in the controls.'.

In spite of the fact that MeSH is a rich resource in biomedical keyword search, it does not contain much information related to disease entity. For this reason our system was low disease recognition. But we should expect more confident recognition when disease recognition is developed by using the UMLS, a resource of the language of biomedicine and health and MeSH is used only search in result.

Level 3 Semantic Interaction

no	HLA	Action	Disease	Sentence
1	A*23, A*26, Cw*04, DRB1*03, DRB1*05	higher	multiple sclerosis (patient, control)	The frequencies of HLA-A23, A26, Cw4, DR3 and, especially DR5 antigens were significantly higher in multiple sclerosis patients than in controls.
2	A*23, DRB1*02	role, development	multiple sclerosis	High prevalence of A23 and DR2 alleles in CDMS patients compared with the normal population may suggest an important role for these alleles in the development of MS.
3	A*23, DRB1*14, DRB1*53	increas	opportunistic infection (patient)	Antigen frequencies in the patients with KS differed from the frequencies in patients with OI by an increase in HLA-A23, -C4, -DR14, and -DR53.
4	A*2301	associat, progress	aids	In contrast, the presence of HLA A*2301 was associated with rapid progression to AIDS, 4% of long term survivors vs. 57.1% of 7 rapid progressors, p = 0.0006, RR = 0.031.
5	A*2301	progress	aids	Influence of HLA alleles on the rate of progression of vertically transmitted HIV infection in children: association of several HLA-DR13 alleles with long-term survivorship and the potential association of HLA-A*2301 with rapid progression to AIDS.
6	A*23187	effect	measles	Calcium ionophore A23187 and phorbol myristate acetate (PMA) were studied for their functional effects on human class-II-restricted measles virus-specific CD4+ cytotoxic T lymphocyte (CTL) clones.
7	A*23187	differen	hiv infection (patient)	Responses of PBL to PMA and A23187 calcium ionophore studied in patients in different stages of HIV infection revealed reduced levels of IL-2 in HIV-infected children beginning before 6 mo of age, and age-dependent increases in expression of IL-4, IL-10, and IFN-gamma.
8	A*375	resist	melanoma, necrosis	After stimulation with lipopolysaccharide and calcium ionophore A23187, culture supernatants of clones c18A and c29A showed cytotoxic activity against human melanoma A375 Met-Mix and other cell lines which were resistant to the tumor necrosis factor, lymphotoxin and interleukin 1.
9	DRB1*02, A*23, B*21	associat	optic neuritis, multiple sclerosis	CONCLUSIONS: The study strongly suggests the association among DR2, A23 and B21 allele and the evolution of ON to MS.
10	DRB1*13	progress, associat	hiv infection	Influence of HLA alleles on the rate of progression of vertically transmitted HIV infection in children: association of several HLA-DR13 alleles with long-term survivorship and the potential association of HLA-A*2301 with rapid progression to AIDS.

Level 2 Semantic Interaction

no	HLA	Action	Disease	Sentence
1	A*23, A*25, A*28, B*04, B*21, B*22, B*27, B*06		bladder tumor (patient)	Statistical analysis showed that HLA antigens A23, A25, A28, BW4, BW21, BW22 and CW4 are more common in bladder tumor patients, whereas B27 and BW6 were more common in the control group.
2	A*23187		death, follicular lymphoma	BACKGROUND: We have previously characterized apoptotic cell death induced in a follicular lymphoma cell line, HF-1, after triggering via the B-cell receptor (BCR) or treatment with Ca(2+) Ionophore A23187.

Fig. 6. Extracted HLA-disease interaction information

Dependency information is an efficient mechanism for extracting relational information from text documents when the entities name listed with and and or. But it is difficult to interpret when HLA entity, disease entity or action keyword are added to the constructed HaD. It is because that we treat HLA and disease into group. We should solve this problem by analysis which it tried every conceivable combination of HLA and disease.

6 Conclusion

In this paper, we proposed the whole procedure to filter short informative sentences automatically using its parse tree. Our procedure consists of two parts: recognizing entities and extracting relation information between entities. In the first part, we efficiently recognized entities by establishing regular expressions and using Mesh ontology. In the second part, we extracted HLA-disease interaction information in sentence of complex structure by searching parse trees. We extracted relation information using 909 abstracts in PubMed and offered the information at our web site. Then, we tested the algorithm with 144 randomly selected sentences. The precision rates reported 89.6% and reported 57.4% in summarization of these sentences. Our algorithm may be extended to other medicine fields such as mental disease and asthma where the relationship between gene and disease is also of importance. We will continue to research an automatic filtering method using machine learning technologies to filter sentences that have no relation between entities without relation and filtering keywords.

References

1. Hanisch, D., Fluck, J., Mevissen, H.-T.: Playing Biologys names Game-Identifying Protein Names in Scientific Text. In: Pacific Symposium on Biocomputing, pp. 403–414 (2003)
2. Hatzivassiloglou, V., Duboue, P.A., Rzhetsky, A.: Disambiguating Proteins, Genes, and RNA in Text - A Machine Learning Approach. Bioinfomatics 1(1), 1–10 (2001)
3. Kazama, J., Makino, T., Ohta, Y., Tsujii, J.: Tuning support vector machines for biomedical named entity recognition. In: Proceedings of the workshop on Natural Language Processing in the Biomedical Domain, July 2002, pp. 1–8 (2002)
4. Friedman, C., Kra, P., Yu, H., Krauthammer, M., Rzhetsky, A.: GENIES: a natural-language processing system for the extraction of molecular pathways from journal articles. Bioinfomatics 17(suppl. 1), S74-S82 (2001)
5. Temkin, J.M., Gilder, M.R.: Extraction of protein interaction information from unstructured text using a context-free grammar. Bioinfomatics 19(16), 2046–2053 (2003)
6. Leroy, G., chen, H., Martinez, J.D.: A shallow parser based on closed-class words to capture relations in biomedical text. Journal of Biomedical Informatics (2003)
7. McDonald, D.M., Chen, H., Su, H., Marshall, B.B.: Extracting gene pathway relations using a hybrid grammar: the Arizona Relation Parser. Bioinfomatics 20(18), 3370–3378 (2004)
8. Horn, F., Lau, A.L., Cohen, F.E.: Automated extraction of mutation data from the literature: application of MuteXt to G protein-coupled receptors and nuclear hormone receptors. Bioinfomatics 20(4), 557–568 (2004)
9. Novichkova, S., Egorov, S., Daraselia, N.: MedScan, a natural language processing engine for MedLine abstract. Bioinfomatics 19(13), 1699–1706 (2003)
10. Schwartz, A.S., Hearst, M.A.: A simple Algorithm for Identifying Abbreviation Definitions in Biomedical Text. In: Pacific Symposium on Biocomputing, vol. 8, pp. 451–462 (2003)
11. Ratnaparkhi, A.: A Maximum Entropy Part-Of-Speech Tagger. In: Proceedings of the Empirical Methods in Natural Language Processing Conference, May 17-18, University of Pennsylvania (1996)
12. Collins, M.: Head-Driven Statistical Models for Natural Language Parsing. PhD Dissertation, University of Pennsylvania (1999)

Description and Verification of Pattern-Based Composition in Coq*

Qiang Liu, Zhongyuan Ynag, and Jinkui Xie

Department of Computer Science and Technology, East China Normal University
Dongchuan Rd. 500, 200241 Shanghai, China
{lqiang,yzyuan,jkxie}@cs.ecnu.edu.cn

Abstract. Design patterns were treated as design components, which serve as elemental components and can be composed to construct a large software system. In the process of composition, the key problem is how to ensure the correctness of composition. In this paper, we use First Order Logic to model some elemental entities and relations in Object Oriented Design, which serve as an ontology in the domain. Then we use the vocabulary in the ontology to specify design patterns and formalize the "faithful" principle as theorems in Coq. Finally, we prove these theorems and show the correctness of composition. As a case study, we described and verified the composition of Composite pattern and Decorator pattern. Once a composition is proven to be correct, one can use the composition repeatedly. This would facilitate reuse of design in a larger scale and reduce errors in design phase.

Keywords: Design patterns, Composition, Verification, Coq, Faithfulness.

1 Introduction

Design patterns [8] are widely used in modern software systems. In the Component-Based approach, design patterns were treated as design components, which serve as elemental components of Object Oriented Design (OOD) and can be composed to construct a large software system. In the process of composition, the key problem is to ensure the correctness of composition. This is a difficult and larger top, and far from being resolved. In this paper, we set our discussion to the domain of the object oriented design, and restrict our discussion in the composition of design patterns. This paper is organized in the following way:

Section 2 introduces related works and comments are also available.

In Section 3, we illustrate the problem by a case study, the composition of Composite and Decorator pattern. In order to model design patterns in Coq, we use First Order Logic to capture the basic entities (Class and Method) and relations in the domain of OOD (Inherit etc). They are both modeled by Inductive types in Coq. These entities and relations are the basic vocabularies based on which we can faithfully describe

---

* The research in this paper was supported by Research Fund for the Doctoral Program of Higher Education of China, No. 20060269002.

T.-h. Kim et al. (Eds.): FGCN 2008 Workshops and Symposia, CCIS 28, pp. 231–245, 2009.

design patterns. The novel thing in our way is that we model design patterns as a list of "propositions", instead of the conjunction of propositions. This can greatly simplify our definitions in latter works. Once we can model design patterns in lists, we can easily describe their composition by concatenating lists. This is the bases for latter works. See Table 3 for details.

In section 4, we give a formal proof of the composition of Composite and Decorator pattern. First, as the composition of design patterns are modeled as lists, the problem of verifying correctness of the composition can be tackled by verifying certain properties of these lists. In this way, we successfully formalized the "faithful" principle as two theorems about "lists" in Coq. Then we devise a couple of customized tactics to facilitate the proof of these theorems. Our tactics take lists (design patterns) as parameters, thus applicable to all the patterns modeled in the correct way (as lists). By using our tactics, we prove that the composition in our example satisfies the properties required by these theorems, which means it is a correct composition in the sense that it is a faithful composition.

Section 5 is conclusion and future works.

To sum up, we have developed a new way of modeling design pattern (as a list of proposition). Based on this model, we can describe their composition as concatenation of lists and formalize properties of compositions as properties of lists, which are easier to operate and verify. In the case study of Composite and Decorator pattern composition, we have shown that the correctness of pattern based composition can be verified with the assist of Coq. Also, by using our carefully designed tactics, the proving processes can be done with a high level of automation.

All the materials presented in this paper are implemented and compiled in Coq v8.1 and available at [14]. Coq [12] is an interactive theorem proving and program development tool, which is widely used in program verification and theorem proving.

2   Related Works

The formalization of design patterns was widely studied [3][5][16], and the works that study their composition are also immense [1][2][4][6][7][15]. [1] studied composition in general, and [2] proposed the concept of 'faithfulness'. Their works can apply to any domain, but in this paper, we only focus on the domain of Object Oriented Design. In [5], the authors formulated basic entities and relations to model the basic concepts in OOD, which is similar to our work, but [5] mainly devoted to graphical representations of OOD to facilitate understanding of system design, and do not considered pattern based compositions. [4][6][15] studied pattern based composition, and used First Order Logic and Temporal Logic of Action to describe design patterns' structure and behavior respectively. Both of them chose the 'faithfulness' principle as correctness criteria, and analyzed the conditions that a correct composition must satisfy. But their proofs are informal and manual, which is based on observations and experiences. We adapt Coq as the proof assistance in this paper, which makes our proof formal and gained a high level of automation.

3 Problem Illustration

In OOD, Class and Method are identified as basic entities, and a set of primary relations were formulated to model the basic concepts in OOD. The number and name of these relations might differ from one another [4][5][6]. In our approach, we define the following inductive type in Coq as in Figure 1.

```
(* Entities as Sets in Coq *)
Parameter  Class : Set.
Parameter  Method : Set.

Inductive  term : Set :=
    (* Class x inherit from Class y *)
| Inherit (x :Class) (y : Class)  : term

    (* Method m is a member function of Class c *)
| Memberfun (m:Method) (c:Class): term

    (* Class x defines a member whose type is reference(s) to instance(s) of Class y *)
| Reference (x:Class)(y:Class):term

    (* Method in Class c1 invokes Method m2 in Class c2 *)
| Invoke(c1:Class)(m1:Method)(c2:Class) (m2:Method) : term

    (*reference of Class c is an argument of Method m *).
| Argument (m:Method)(c:Class):term
```

Fig. 1. Inductive definition of term

Words in (* *) are comments. This definition serves as a grammar of writing specifications about design patterns. Some authors [5] treat entities as unary relations. However, we just define them as Sets in Coq. Treating them in this way gives us a typed system, which improves readability and simplicity. As we can see from the inductive definition of term, all the relations are typed, taking parameters from Class and Method.

The clauses are actually the constructor of the Set term. It is easy to add new relations by adding new clauses in the inductive definition. Our definition of term is flexible enough to add any relations that character the structure of design patterns. Note we only focus on the structural aspect of design patterns in this paper. Though our definition is general enough to model patterns in the structural category in [8], it should be extended to model the behavioral aspect of design pattern. We leave it as our future work, see Section 5 for details.

Usually, it will take many terms to precisely describe a design patter. So we model design patterns as a list of 'terms':

$$\text{Definition DesignPattern:= list term.} \tag{1}$$

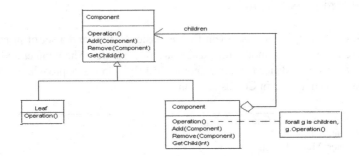

Fig. 2. Composite pattern in UML

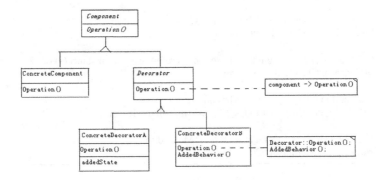

Fig. 3. Decorator pattern in UML

The benefit of this definition is fully discussed in Section 1. Now, as a case study, consider the composition of Composite pattern and Decorator pattern. Their descriptions in UML class diagram are presented in Figure 2 and 3.

By the definition in Figure 1 and Formula (1), we can formally define Composite pattern and Decorator pattern by using a list of 'terms' as in Figure 4 and 5. Figure 4 defines the participants in Composite pattern one by one. Note in the definition of Reference relation, we do not explicitly state the name of the instance of Class component and its cardinality. The reason for this is that our way is based on Class and Method, we do not want to include the Object into our model, which is related to the behavior of design patterns. As one can expected, in the definition of the Invocate relation, we also omit the information related to Object and only specify the Type of the Object, i.e., the Class. However, as in [5], this can be handled by introducing more relations between Class and Method, or even between the sets of them. This is not a problem for our model, because, as mentioned above, we can easily introduce new relations by adding new clauses to the inductive definition of 'term'. However, the focus in our paper is not to complete the set of relations, rather, we focus on introducing the approach of verifying pattern-based composition in Coq.

```
Definition composite_pattern : DesignPattern :=
(* Define the interface for the Class component *)
  Memberfun operation component
::Memberfun add component
::Memberfun remove component
::Memberfun getchild component

(* Define the interface for the Class composite *)
::Memberfun operation composite
::Memberfun add composite
::Memberfun remove composite
::Memberfun getchild composite

(* Class composite references instances of Class component *)
::Reference composite component

(* Method invocation *)
::Invocate composite operation component operation

(* Define the interface of Class leaf *)
::Memberfun operation leaf

(* Define parameters of each Method *)
::Argument add component
::Argument remove component
::Argument getchild int

(* Define the hierarchy of Class Inheritance *)
::Inherit leaf component
::Inherit composite component::nil.
```

Fig. 4. Definition of Composite pattern in Coq

In practice, when software designers want to compose these two patterns, depending on their experiences and intuitions, they may compose them as in Figure 6 (in UML class diagram). In fact, this composition forms as a new pattern. We call it the Composite_Decorator pattern. Its formal counterpart is in Figure 7.

However, we cannot rely on the experiences and intuitions of the designer, because this work is laborious and error-prone. First, there are many statements in each pattern, the designer may lose statements in the original patterns, or add undesirable new statements in the composition. Additionally, as we can see in Figure 6, there is a mapping between entities in the separate design patterns before the composition and entities in after the composition, because some entities play roles in both separate patterns, for example, the class "TheComponent", the method "TheOperation". Therefore, we need to define functions to record all these information for the designer, which serve as a base for the proof of correctness of pattern compositions. That is the main job in the next section.

Definition decorator_pattern : DesignPattern :=
(* Define the interface of Class component, concretecomponent and decorator *)
 Memberfun operation component
::Memberfun operation concretecomponent
::Memberfun operation decorator

(* Method invocations *)
::Invocate decorator operation component operation
::Invocate concretedecoratorb operation decorator operation
::Invocate concretedecoratorb operation concretedecoratorb addedbehavior

(* Define the interface of Class concrete decorators *)
::Memberfun operation concretedecoratora
::Memberfun operation concretedecoratorb
::Memberfun addedbehavior concretedecoratorb

(* References in this pattern *)
::Reference concretedecoratora addedstate
::Reference decorator component

(* Define the hierarchy of Class Inheritance *)
::Inherit concretecomponent component
::Inherit decorator component
::Inherit concretedecoratora decorator
::Inherit concretedecoratorb decorator::nil.

Fig. 5. Definition of Decorator pattern in Coq

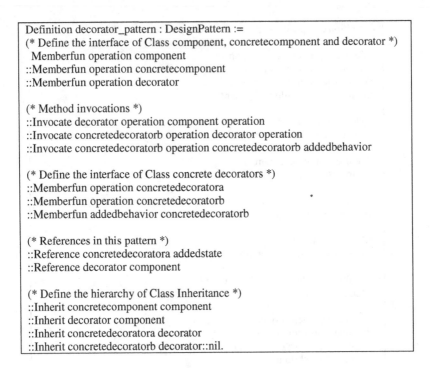

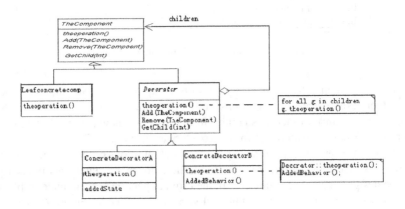

Fig. 6. The composition of Composite and Decorator in UML

4 The Formal Proof

This Section is the main technical part of this paper. Firstly, Section 4.1 introduces the correctness criterion. Then in Section 4.2 the mapping functions are defined for the case study and the correctness criterion is formalized in Coq. Finally, in Section 4.3,

```
Definition  composite_decorator_composition: DesignPattern :=
(* Define the interface of Class thecomponent *)
  Memberfun theoperation thecomponent
::Memberfun add thecomponent
::Memberfun remove thecomponent
::Memberfun getchild thecomponent

(* Define the interface of Class leafconcretecomp *)
::Memberfun theoperation leafconcretecomp

(* Define the interface of Class compositedecorator *)
::Memberfun theoperation compositedecorator
::Memberfun add compositedecorator
::Memberfun remove compositedecorator
::Memberfun getchild compositedecorator

(* References in this pattern *)
::Reference compositedecorator thecomponent
::Reference concretedecoratora addedstate

(* Define the interface of concrete decorators *)
::Memberfun theoperation concretedecoratora
::Memberfun theoperation concretedecoratorb
::Memberfun addedbehavior concretedecoratorb

(* Method invocation in this pattern *)
:: Invoke concretedecoratorb theoperation decorator operation
::Invoke concretedecoratorb theoperation concretedecoratorb addedbehavior

(* Arguments of each method *)
::Argument add thecomponent
::Argument remove thecomponent
::Argument getchild int

(* Define the hierarchy of Class Inheritance *)
::Inherit leafconcretecomp thecomponent
::Inherit compositedecorator thecomponent
::Inherit concretedecoratora compositedecorator
::Inherit concretedecoratorb compositedecorator::nil.
```

Fig. 7. Definition of composition of Composite and Decorator in Coq

we finished the proofs of the theorems formalized in Section 4.2 and defined two customized tactics to allow some automation in the proving processes.

4.1 The Correctness Criterion

To prove the correctness of composition, we first need a correctness criterion. We choose the well known principle of 'faithfulness' as our correctness criterion. In general, for a composition to be faithful, it must satisfy the follow two conditions:

No component loses any properties after composition.

No new properties about each component can be added after composition.

Particularly, as to our case study, this means: for the composition of Composite and Decorator pattern to be correct, it must satisfy the following two conditions:

No statement in Composite and Decorator loses after the composition.

No new properties about Composite and Decorator can be added after the composition.

4.2 Formalizing the Correctness Criterion in Coq

Before formalizing the correctness criterion, we should define some mapping functions between entities (Class and Method) before and after the composition. These functions contain the key information of the composition. Now we explicitly define these mapping functions in Figure 8.

```
(* Mapping from Class to Class *)
CM : Class -> Class
(* Mapping from Method to Method *)
MM :Method ->Method
(* Mapping from Composite to Composite_Decorator *)
Mcmp:Composite->Composite_Decorator
(* Mapping from Decorator to Composite Decorator *)
Mdcr:Decorator->Composite_Decorator
(* The whole Mapping relation is the combination of Mcmp and Mdcr *)
Mcmp_dcr := Mcmp V Mdcr.


(* Concrete mapping functions *)
(* Class mapping from Composite pattern to the composition  *)
Definition ClsMcmp : Class -> Class := fun c:Class => match c with
| component=>thecomponent
| leaf =>leafconcretecomp
| composite =>compositedecorator
| others => others   end.
(* Method mapping from Composite pattern to the composition *)
Definition MtdMcmp : Method -> Method := fun m => match m with
| operation => theoperation
| others => others   end.
(* Class mapping from Decorator pattern to the composition *)
Definition ClsMdcr : Class -> Class :=fun c:Class=>match c with
| component => thecomponent
| concretecomponent => leafconcretecomp
| decorator => compositedecorator
| others => others   end.
(* Method mapping from Decorator pattern to the composition *)
Definition MtdMdcr : Method -> Method :=fun m => match m with
| operation => theoperation
| others=> others   end.
```

Fig. 8. Mapping functions in Coq

Mcmp and Mdcr are maps defined based on CM and MM. For details of these definitions, see [14].

Beasd on these definitions, we now formalize the principle of faithfulness in Coq. As mentioned in Section 1, the theorems are actually properties about lists, which is easier to formulate and verify in Coq.

$$\forall (\text{ t:term}), \text{In t (Composite} \lor \text{Decorator)} \rightarrow$$
$$\exists (\text{s:term}), \text{In s Composite\_Decorator} \land (\text{Mcmp\_dcr t=s}). \qquad (2)$$

$$\forall (\text{ t:term}), \text{In t Composite\_Decorator} \rightarrow$$
$$\exists (\text{s:term}), \text{In s (Composite} \lor \text{Decorator)} \land (\text{Mcmp\_dcr s=t}). \qquad (3)$$

(Composite $\lor$ Decorator) is not only the disjoint union of Composite and Decorator, it is the deductive closure of their union. The closure is calculated by our function "Union" and "Closure" in Coq, see Figure 9 and 10 for details.

Usually, we allow some new facts to be deduced from the union of two patterns, with explicitly specified rules. Otherwise, any other new facts would be undesirable, and would be regarded as an error in the composition. As far as the structure is concerned, we would like to define some larger building block in the OOD, for instance, the polymorphism, which can be described by the elemental constructs in the inductive type 'term'. Clearly, this related to a more sophisticated problem, that is: what properties should remain after the composition. For simplicity, we just concern the elemental statements in each pattern in this paper. However, to make our approach more practical for everyday use, we should concern more complex properties in each pattern. We leave it as our future work.

In our example, there is only one rule allowed for reasoning, that is:

$$\text{Inherit a b, Inherit b c} \Rightarrow \text{Inherit a c.} \qquad (4)$$

```
(* Inductively calculate the Union of list l and m, which remove the same elements
in l and m *)
Fixpoint union (l m:list term) { struct l } : list term:=
match l with

(* if l is null, then return list m *)
| nil => m

(* if l is not null, check the number of occurrence of a in m *)
| a::l' =>
if Z_eq_dec (nb_occ a m) 0

(* if a does not occur in m, then add a to the resulting list *)
then  a::union l' m

(* if a occur in m, then do not add a to the resulting list *)
else union l' m end.
```

Fig. 9. Definion of Union

Other rules can easily be added to our definition ('fsttrans') by simply adding a new clause in its definition. Using the definition of closure, we can now calculate the list of terms before and after the mapping, the composite_decorator and the composite_decorator' respectively. See Figure 11.

4.3 Proving Correctness of the Composition

To facilitate the proof of these theorems, we also need to define some customized tactics. These tactics allow some automation in the proving process. Tactics are powerful tools in Coq, which facilitate the proving process of the user. There is a language for the definition of tactics in Coq. In the proofs-as-programs diagram, this language is rather for the user to write down the proving process for their specific problems. In our case, we should use this language to automate most of the task involved in proving the Theorems.

We now define the tactic not_add_tac for the proof of theorem (2) and not_lost_tac for the proof of theorem (3). See Figure 12 and 13 for the definition of these tactics.

```
(* Define a function which calculate the transitive clouse of term t in list l *)

Fixpoint fsttrans (t:term) (l : list term) {struct l} : list term :=
match l with

(* if l is nil, return null *)
| nil => nil
| a::l' =>

(* match t and the first element in list l,if a and t can trigger the rule in (4), then
and the derived term the the resulting list  *)
    match a, t with
    | Inherit f1 f2, Inherit s2 s3 =>
     if ceqbool f2 s2 then Inherit f1 s3::(fsttrans t l')  else
     if ceqbool s3 f1 then Inherit s2 f2::(fsttrans t l')  else
     fsttrans t l'

(* if rule (4) cannot be applied, then check the next element in list l. *)
| _ , _ => fsttrans t l'
    end
end.

(* Based on the definition of fsttrans, inductively calculate the deductive clouse of
list l. *)
Fixpoint closure (l :list term) { struct l } : list term:=
    match l with
     | nil => nil
     | a::l' =>a::(fsttrans a l')++(closure l')
    end.
```

Fig. 10. Definition of Closure

```
(* The composite_decorator pattern before the composition *)
Definition composite_decorator :=
            closure (union composite_pattern decorator_pattern).
   (* The composite_decorator' pattern after the composition *)
Definition composite_decorator' :=
            closure (union composite_pattern' decorator_pattern').
```

Fig. 11. Defintion of patterns before and after the composition

```
(* This tactic takes list l1 and l2 as parameters, which means it is suitable for
   any patterns, as long as they are modeled in our way---as a list of 'terms' *)

Ltac not_add_tac l1 l2 :=
(* First use the atom tactic intro provided by the Coq system. This also allows us
   to use all the conditions available to the theorem. *)
   intro t;

(* The atom tactic simpl automatically simplify the theorem according to the defi-
nitions in the user's environment. *)
   simpl;

(* Intuition is a powerful tactic in Coq, it analyzes the current situation and provide
subgoals whenever necessary. *)
intuition;
match goal with

(* Match the form of the current goal, according our definition of Mapping func-
tions, we can automate the proof of theorem (2). *)
| [id1 : Mcmp ?x =t |- _ ] => exists x
| [id2 : Mdcr ?x=t |- _ ] => exists x
| [id3 : Mcmp_dcr ?x = t |- _ ]=> exists x
| [id4 : _ |- _ ] => auto
end;
auto.
```

Fig. 12. Definition of the not_add_tac tactic

After these general definitions of tactics, we are now ready for the proof of theorem (2) and (3). Figure 14 and 15 are the details of the proving processes. Note in Figure 14, theorem (2) is done with only one tactic: the not_lost_tac. We only need to pass the correct actual parameters to the not_lose_tac tactic and then the Coq system will automatically search the solution to the goal based on the user's definitions in the current environment.

Note in Figure 15, the tactic not_add_tac has done most of the job in the proof of theorem (3). According to our definition of the not_add tac, the Coq system will automatically search the right terms for the 'exists' statements in the goal. Howerver, as we explicitly allow the deductive rule in formula (4), thus there are some extra lines in the proof, which need to be handled manually and this is only a tiny part of the whole proof.

Till now, we have demonstrated our formal way step by step, and have proven that the example composition is a correct one. By following our way, designers can prove the correctness of their own composition. Firstly, the designer should translate the informal description of design patterns in UML to the formal description in Coq. This task is laborious and we should develop some tools to automate this translation. Then, as we can see, the definition of the type of mapping functions and the parameterized tactics are general enough for the designer to reuse. The only information that designer need to offer is the implementation of the mapping functions, as we did in the examples in Figure 8, which are specific to the Composite and Decorator pattern. After implementing the mapping functions, the way followed is rather straightforward. As we model design patterns as a list of 'terms', the formalization of the correctness criterion for designer's own patterns can be obtained by just replacing the name of the lists occurred in formula (2) and (3). Finally, the proving process is rather easy by passing the correct parameters to our customized tactics.

Once a composition is proven to be correct, it can be reused repeatedly in the process of system construction. This would facilitate reuse of design in a larger scale and reduce errors in design phase, which justifies all the efforts of verifying their correctness.

(*This tactic takes list l1.l2,M as parameters,which is more general than not_add_tac, reader may redefine the not_add_tac to make it more general, however the optimization of that tactic is not in the interest of this paper *)

Ltac not_lose_tac l1 l2 M:=

(* Match the current goal. If the conclusion in the current goal is of the form after ⊢, then we can actually 'program' the proof according the values of parameters, i.e., l1,l2 and M. *)

match goal with

| [⊢forall t:term, In t l1 ->(exists s:term, In s l2∧M t=s)]=>

(intro t ;

  simpl;

  intuition;

(* the term t is the term we need to find in l1. Usually this is done manually, however our tactic help us find the right term in l1. *)

  exists (M t);

  subst; simpl; auto.)

end.

Fig. 13. Definition of the not_lose_tac tactic

> (*This theorem shows that for any desired property (term) of the compos-
> ite_docorator, there is a correspondence in the composite_decorator', that is no prop-
> erties are lost. *)
>
> Theorem not_lose_Mcmp_dcr :
> forall(t:term), Int composite_decorator ->
>   (existss:term, In s composite_decorator'∧Mcmp_dcr t=s).
> Proof.
> (*Pass the composite_decorator, composite_decorator' and Mcmp_dcr to tactic. *)
> not_lose_tac composite_decorator composite_decorator' Mcmp_dcr.
> Qed.

Fig. 14. The proof of the theorem not_lose_Mcmp_dcr

> (*This theorem shows that for any properties in the composition, there is a corre-
> spondence in the original patterns, that is, no undesirable properties are added. *)
>
> Theorem not_add_Mcmp_dcr : forall (t:term), In t composite_decorator' ->
>   (exists s:term, In s composite_decorator∧Mcmp_dcr s=t).
> Proof.
>   not_add_tac composite_decorator' composite_decorator.
>   exists (Inherit concretedecoratora component). auto .
>   exists (Inherit concretedecoratorb component).auto.
> Qed.

Fig. 15. The proof of the theorem not_add_Mcmp_dcr

5   Conclusions and Future Work

The main contribution of this paper is introducing a new approach for the formal
verification of pattern-based composition in Coq. We model the OOD in an inductive
type 'term' with several elemental constructs which capture basic concepts in the
domain of OOD. Then we defined some mapping functions and customized tactics to
facilitate the proving process. By using our basic definitions, designers can conven-
iently formulate desirable properties of design patterns as theorems in Coq and then
prove their correctness. Once a composition is proven to be correct, it can be reused
without any further proofs. This will facilitate reuse in a larger scale (not only reuse
of design pattern, but also their composition), which can benefit many aspects in the
developing process, including developing efficiency (using larger building blocks)
and reduce errors in design phase (the composition is proven to be correct). Additionally, the composition is a larger logic unit (like the Composite_Decorator pattern),
which makes the system design easier to understand and maintain.

However, in this paper, our work is rather primitive. We only focus on the structural aspect of design patterns. As we all know, design patterns have both structure
and behavior. Though for some patterns it is enough to focus on their structural aspects, some design patterns do have behavioral significance, like Observer pattern. As
we can see, the elemental constructs in the definition of 'term' are all about Class and

Method. These relations are static relations, which, once coded in the program, cannot be changed during the run time. Clearly, these relations are just suitable for modeling the structural aspect of design patterns. To model behavioral aspects of design patterns, we should concern relations about Objects. Many authors [4][6] use temporal logic, like Temporal Logic of Action [9], to formalize the behavioral aspect of design patterns. It is our future concern to integrate these two aspects in Coq.

Furthermore, to achieve a higher level of automation in the proving of composition, we should develop a GUI tool to convert UML diagrams to Coq scripts. Usually, the graphical representation of design patterns in UML can be easily exported to XMI format. Then we need to convert this format to the scripts in Coq. This would hide details about Coq from designers, thus improve the work efficiency of system designers.

References

1. Abadi, M., Lamport, L.: Composing specifications. ACM Transactions on Programming Languages and Systems 15(1), 73–132 (1993)
2. Moriconi, M., Qian, X., Riemenschneider, R.A.: Correct architecture refinement. IEEE Transactions on Software Engineering 21(4), 356–372 (1995)
3. Tommi, M.: Formalizing design pattern. In: Proceedings of the 20th International Conference on Software Engineering, pp. 115–124 (1998)
4. Taibi, T., Ngo, D.C.L.: Formal specification of design pattern combination using BPSL. International Journal of Information and Software Technology (IST) 45(3), 157–170 (2003)
5. Eden, A.H., Hirshfeld, Y.: Principles in formal specification of object-oriented architectures. In: Proceedings of the 11th CASCON, Toronto, Canada (November 2001)
6. Dong, J., Alencar, P., Cowan, D.: Ensuring structure and behavior correctness in design composition. In: Proceedings of the 7th Annual IEEE International Conference and Workshop on Engineering of Computer Based Systems (ECBS), Edinburgh UK, pp. 279–287 (2000)
7. Keller, R., Reinhard, S.: Design components: towards software composition at the design level. In: Proceedings of the 20th International Conference on Software Engineering, pp. 302–311 (1998)
8. Gamma, E., Helm, R., Johnson, R., Vlissides, J.: Design Patterns, Elements of Reusable Object-Oriented Software. Addison-Wesley Publishing Company, Reading (1995)
9. Lamport, L.: The temporal logic of actions. ACM Transactions on Programming Languages and Systems 16(3), 873–923 (1994)
10. Coquand, T., Huet, G.: The calculus of constructions.Technical Report 530, INRIA (1986)
11. Girard, J.-Y., Lafont, Y., Taylor, P.: Proofs and Types. Cambridge Tracts in Theoretical Computer Science, vol. 7. Cambridge University Press, Cambridge (1989)
12. Bertot, Y., Castéran, P.: Interactive Theorem Proving and Program Development. Coq'Art: The Calculus of Inductive Constructions. In: Texts in Theoretical Computer Science. Springer, Heidelberg (2004)
13. Alencar, P., Cowan, D., Lucena, C.: A formal approach to architectural design patterns. In: Gaudel, M.-C., Woodcock, J.C.P. (eds.) FME 1996. LNCS, vol. 1051, pp. 576–594. Springer, Heidelberg (1996)

14. Pattern Composition in Coq,
http://ecnucs-selab.googlecode.com/files/exampleinpaper
15. Dong, J., Alencar, P.S.C., Cowan, D.D., Yang, S.: Composingpattern-based components and verifying correctness. The Journal of Systems and Software, 1755–1769 (2007)
16. Eden, A.H., Hirshfeld, Y., Kazman, R.: Abstraction classes in software design. IEE Software 153(4), 163–182 (2006)

A Synthetic Reverse Model Based on AIS and the Application

Chang-You Zhang[1,2], Wen-Qing Li[2], Yuan-Da Cao[2], and Zeng-Lu Fan[3]

[1] School of Computer & Information Engineering, Shijiazhuang Railway Institute,
Shijiazhuang, Hebei, 050043, P.R. China
[2] School of Computer Science & Technology, Beijing Institute of Technology,
Beijing, 100081, P.R. China
[3] Hebei Meteorology Bureau, Shijiazhuang, Hebei, 050043, P.R. China
zhangchangyou@tsinghua.org.cn, {liwq,caoyd}@bit.edu.cn,
zl_fan@yahoo.com.cn

Abstract. Artificial Immune System (AIS) is a set of principles and algorithm following the properties of nature immune system. In order to improve the performance of IDS, a synthetic dimension reduction model is proposed in this paper. First of all, we define a similarity distance algorithm between two vectors based on analogy reasoning. Then, we introduce an optimization method to meliorate the normal-behavior-set and abnormal-behavior-set based on AIS and Genetic Algorithm (GA). And then, we construct a synthetic reverse model taking both of the above mentioned behaviour set. When a new behavior sample is sniffered from network, the distances between this behavior and each of the two meliorated sets are calculated. Finally, we treat these two distances as ordinate and abscissa, the new behavior sample is mapped from a multi-dimensional vector space into a point in a two-dimensional coordinate plane. According to the position of this point, we determine whether a behavior is an intrusion or not.

Keywords: Synthetic Reverse Model, Artificial Immune System, IDS.

1 Introduction

Biology system is a fountain where people continually abstract inspirations. Artificial Immune System (AIS) is one of these main inspirations. The mathematic framework of the immunity network was raised by Jerne[1] in 1970s. In 1986, Farmer constructed a dynamic model for immune system based on the mathematic framework of the immunity network. The following researchers raised some new theories from different viewpoints of the biology immune system to perfect the AIS model, algorithms and applications. The AIS shows its powerful abilities in information processing and problem resolving in many fields, especially in information security, pattern recognition, intelligent optimization, machine learning, data mining, robotics, diagnostics and cybernetics etc[2].

Intrusion Detection System (IDS) is a very popular security defense. The aim of AIS is to monitor, detect and identify a baleful or attempted behavior to network and computer system by means of collecting and analyzing the system information. Thus,

T.-h. Kim et al. (Eds.): FGCN 2008 Workshops and Symposia, CCIS 28, pp. 246–257, 2009.

IDS can distinguish whether a system state is "normal" or "abnomal"[3]. So, an IDS can be defined as the guard system that automatically detects malicious activities within a host or a network, and consequently generates an alarm to alert the security apparatus at a location if intrusions are considered to be illegal on that host or network [4]. Intrusion Detection System is divided into two categories: anomaly detection and misuse detection [5].In the requirements of an IDS, correctness and real-time are two important items. In high-speed network, IDS faces a major problem that the low detection rate can not handle the massive data transmission. There are two basic ideas to solve this problem: (1) improve processing capabilities of the intrusion detection system, including data-processing capacity and data collection capacity. (2) introduce new algorithms or pretreatment to reduce the complexity of data processing.

This paper is organized as follows: in section 2, the background and related works are reviewed. Then a behavior set evolution scheme and a similarity distance of network behavior is proposed in section 3. Section 4 introduces a principle and the framework of the synthetic reverse model. Section 5 explains the dimension reduction method of behavior samples. Finally, it goes to conclusions.

2 Related Works

Accordance with the second view mentioned above, we improve the efficiency of data processing by means of reduce the behavior vector dimensions. Manifold learning is a way of simplifying high-dimension data by finding low-dimensional structure in it. The goal of the algorithms is to map a given set of high-dimensional data points into a surrogate low-dimensional space [6]. Animesh Patcha [7] presented an anomaly detection scheme, called SCAN (Stochastic Clustering Algorithm for Network Anomaly Detection), that has the capability to detect intrusions with high accuracy even with incomplete audit data.

Many researchers proposed some new intrusion detection methods oriented new network entironments. Paper [8] proposed an Anomaly detection enhanced classification in computer intrusion detection. Kim [9] proposes a method of applying Support Vector Machines to network-based Intrusion Detection System (SVM IDS). Park [10] proposes a new approach to model lightweight Intrusion Detection System (IDS) based on a new feature selection approach named Correlation-based Hybrid Feature Selection (CBHFS) which is able to significantly decrease training and testing times while retaining high detection rates with low false positives rates as well as stable feature selection results. Taylor [11] introduced an lightweight IDS solution called NATE, Network Analysis of Anomalous Traffic Events. The approach features minimal network traffic measurement, an anomaly-based detection method, and a limited attack scope. Horng [12] presents the results of a study on intrusion detection on IIS (Internet information services) utilizing a hybrid intrusion detection system. The feasibility of the hybrid IDS is validated based on the Internet scanner system (ISS).

Biology immune system has some mystic abilities in recognition, learning and memory. These abilities are of the characters of distribution, self-organization and diversity. The main AIS principles consist of immune recognition, immune learning, immune memory, clone selection, diversity generation and maintenance etc. It is generally deemed that the former four principles form the immune response process of AIS.

The network-based IDS highly resemble AIS in their protecting functions. Kim studied on the intrusion detection based on clone selection and negative selection mechanism. The experiment results which he provided proved that using a negative detection operator was a key factor to keep a low false positive rate. The antibody take the responsibility of detecting pathogens, therefore the creation, evolvement and how it work are the key steps in modeling AIS. So, how to simulate the updating of the gene set, the negative selection, the clone selection are the vital job to establish an intrusion detection system in terms of the AIS principles. Table 1 list the comparison on conceptions in AIS and IDS.

Table 1. Comparison on Conceptions in AIS and IDS

AIS	IDS
Antibody	Checking pattern
Antigen	Nonself pattern
Binding	Checking pattern matching with the nonself pattern
Negative selection	Negative selection
Lymphocyte clone	Monitor copy
Antigen monitoring	The monitoring of IDS
Clearing antigen	Monitor response

3   Similarity Distance of Network Behavior

3.1   Behavior Modeling

In this paper, a network behavior is described in a vector with some attributes in high correlation. Generally, these attributes mainly include: service types (*srvType*), source address (*srcIP*), source port (*srvPort*), destination address (*dstIP*), destination port (*dstPort*), duration (*dur*), number of bytes sent from Source port (*srcBytes*), number of bytes sent to destination port (*dstBytes*), state (*flag*). Therefore, each of the network behavior is expressed by a 9-dimensional vector as equation (1),

$$\vec{X} = [srvType, srcIP, srvPort, dstIP, dstPort, dur, srcBytes, dstBytes, flag]^T \tag{1}$$

According to data types, these elements of vector $\vec{X}$ can be sorted into two parts: (1) Character elements. The result of their matching operation is just a TRUE or FALSE. Such elements are suitable for analogy reasoning algorithm that will be talked in this paper. This type of elements are service type (*srvType*), source address (*srcIP*), source port (*srvPort*), destination address (*dstIP*), destination port (*dstPort*), state (flag), and so on. (2) Numerical elements. The value of such element is a number. The result of comparing operation between them is a number too, not a boolean value. These elements do not fit for the formula which is proposed in this paper. They are delay (*dur*), number of bytes sent from Source (*srcBytes*), number of bytes sent to destination (*dstBytes*), and so on.

The numerical elements should be preprocessed to be appropriate for the synthetic dimension reduction method. The detail of the discretization algorithm can be refered to paper [13]. After discretization, the numerical element is turned into a character one, and finally, the vector, $\vec{X}$, will be fit for the algrithms of similar distance calculation.

3.2 The Similar Distance Algorithm

In this paper, we regard the similarity between two network behaviors as the base to establish an intrusion detection system. The similarity algorithm grounds on a synthetic dimension reduction model in analogy reasoning. Both the contribution of the similar elements and the dissimilar elements are taken into consider for the similarity measurement.

Here, the N-dimensions vector (2),

$$\vec{X} = [x_1, x_2, \cdots, x_n],\tag{2}$$

represents a network behavior. The x_i ($i = 1,\ldots,n$), is one of the elements in the vector $\vec{X}$, standing for an attribute of the network behavior. The similarity between two behaviors shows their likeness degree. For the convenience, some definitions are listed as follows.

Definition 1. Similarity between two Behavior Vectors

$\vec{X}$ and $\vec{Y}$ stand for two behavior vectors respectively, the similarity between them can be calculated by means of formula (3).

$$Sim(\vec{X},\vec{Y}) = \frac{f(\vec{X} \cap \vec{Y})}{f(\vec{X} \cap \vec{Y}) + \alpha \bullet f(\vec{X} - \vec{Y})}, (\alpha \geq 0)\tag{3}$$

In formula (3), function $f(\vec{X} \cap \vec{Y})$ denotes the similarity contribution of the matched elements, and the function $f(\vec{X} - \vec{Y})$ denotes similarity contribution of the dismatched elements. α is a coefficient for the contribution of the dismatched elements. The value of α is not less than 0. Evidently, the value of $Sim(\vec{X},\vec{Y})$ is larger than 0 and less than 1.

Definition 2. Similarity between a Behavior Vectors and a Behavior Set

If set A is a behavior set, and $\vec{X}$ is a behavior vector, then the similarity between them can be calculated by means of formula (4).

$$Sim(\vec{X},A) = \max\{Sim(\vec{X},\vec{A_j}), \vec{A_j} \in A, j = 0,1,\cdots,m\}\tag{4}$$

In formula (4), $Sim(\overrightarrow{X}, \overrightarrow{A_j})$ is the similarity between $\overrightarrow{X}$ and $\overrightarrow{A_j}$, the element j in set A. The final result will be equal to the maximum one.

3.3   The Sample Set Optimization Based on Artificial Immune

Artificial Immune System imitates the natural immune system, and provides a novel way to solve the problem of potential. To avoid the training data sets may be some faulty, we use the artificial immune and genetic algorithm method to optimize the samples set of abnormal behavior AI_0. The optimization process is showed in Figure 1.

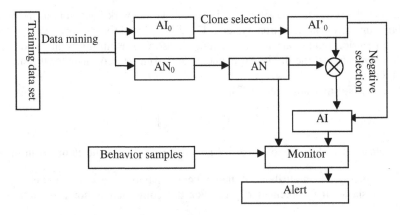

Fig. 1. Sample set creating and optimizing processes

Figure 1 mainly states the generation and optimization process of abnormal behavior set AI. Firstly, data mining method is used in generating initial sample set AI_0. And this result can be supplemented based on experience. Then, the genetic mutation and proliferation are operated on AI_0 to generate a larger candidate sample set AI_0'. And then, the affinity of every sample will be calculated by means of the similarity to AI_0. The samples with high affinity will be selected. And then, the further negative selection delete the samples with the simlarity to AN equal to 1 or very close to 1. Finally, we get a optimized sample set of abnormal behavior. The optimization is of a two-step process:

(1)Clone Selection

The purpose of clone selection is to expand the quantity of abnormal samples, or to optimize the distribution characteristic of the sample space. These characteristic include the distribution density, proportional spacing of the samples, etc.. In this section, our cloning algorithm take AI_0 as original parameter, and adopt multi-crossing and random variation to expand sample space and enhance the homogenization of the spatial distribution. Expanding the abnormal behavior sample space and optimizing the homogenization of sample distribution are propitious to reduce the rate of false negative alerts in detection system.

The kernel idea of the clone selection algorithm is to measure the similarity between sample $\overrightarrow{X}$ and AI_0. The main steps showed in Figure 2.

Step1: Obtain AI_0 by means of data mining;

Step2: Create AI_0 (t_1) through muti-point intercross;

Step3: Construct AI_0 (t_2) after random variation;

Step4: Set the affinity threshold v, and let $AI_0 = AI_0$ (t_2) ;

Step5: Suppose the element number of AI_0 is M, repeat operating on $\overrightarrow{Xi} \in AI_0$, and let $i = 0$;

Step6: Calculate the similarity (fi) between $\overrightarrow{Xi}$ and AI_0;

Step7: Judge on $fi > v$, $true \Rightarrow$ reserving, $fi = 1$ or $false \Rightarrow$ deleting, $i++$;

Step8: Judge on $i < M$, $false \Rightarrow$ goto Step5; $true \Rightarrow$ goto Step9;

Step9: Construct AI_0 (t_3) , finish.

Fig. 2. Clone selection algorithm

In figure 2, we let the affinity of $\overrightarrow{Xi}$ and AI_0 equal to the similarity between $\overrightarrow{Xi}$ and AI_0, i.e. $fi = S_{AI}^X$. When the value of fi exceed the given threshold v, the sample $\overrightarrow{Xi}$ will be reserved in AI_0, else it will be deleted. There is a special case that the $fi = 1$. That means $\overrightarrow{Xi}$ is a duplicate sample, and it is not necessary for AI_0. Therefore we delete it. Finally, we get AI_0 (t_3) .

Step1: Read in AI_0 (t_3) , $AN = AN_0$ as parameters, set affinity threshold v;

Step2: Suppose the element number of AI_0 (t_3) is N, repeat operating on $\overrightarrow{Xi} \in AI_0(t_3)$, and let $i = 0$;

Step3: Calculate the similarity between $\overrightarrow{Xi}$ and AN. Let affinity $fi = S_{AN}^X$;

Step4: Judge on $fi > v$, $true \Rightarrow$ deleting, $false \Rightarrow$ reserving, $i++$;

Step5: Judge on $i < N$, $false \Rightarrow$ goto Step4; $true \Rightarrow$ goto Step7 ;

Step7: Construct AI, finish.

Fig. 3. Negative selection algorithm

(2)Negative Selection

In immune system, negative selection is to protect body cells from mis-damage. In other words, there can not exist any behavior vector in AI that is same or very similar to ones in AN. In negative selection algorithms, the similarity between a element in AI and the behavior set AN need to be calculated. Based on this similarity, the elements with a similarity equal to 1 or very close to 1 will be ruled out to avoid false positive alerts. The steps are listed in Figure 3.

4   Synthesized Distance Model

4.1   Mathematic Modeling

As we described above, a network behave was abstracted into an n-dimensional vector, $\vec{X} = [x_1, x_2, \cdots, x_n]$. It is called as a point in n-dimensional vector space. The collection of all these points constitutes a behavior surface. In intrusion detection system, we are concerned about abnormal behavior and normal behavior of collections. In order to facilitate drawing, a collection of normal behaviors will be abstracted to a "normal plane" in a three-dimensional space, and a collection of abnormal behaviors will be abstract to a "abnormal plane". These two planes are shown in Figure 4 and Figure 5. Here, P and Q represent two points in this space.

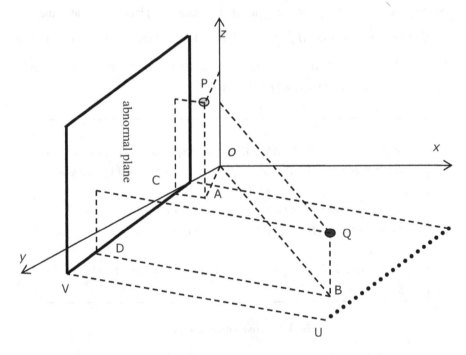

Fig. 4. The simple distance model of behavior vector to abnormal plane

In Figure 4,

> | AC | is the distance from point P to the "abnormal plane".

> | BD | is the distance from point Q to the "abnormal plane ";

> | UV | is the threshold.

∵ | AC | < | UV |, and | BD | < | UV |;

∴ P and Q are abnormal.

Let $fp\ (x)$ is the abnormal probability function of x,

then $fp\ (P) =$ | UV | - | AC | and $fp\ (Q) =$ | UV | - | BD | represent the abnormal probability of P and Q. And then,

∵ | AC | < | BD |,

∴ $fp\ (P) > fp\ (Q)$.

That is to say the abnormal probability of point P is higher than that of Q.

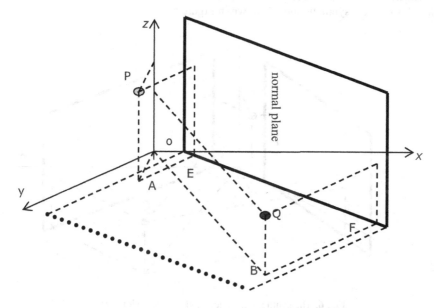

Fig. 5. The simple distance model of behavior vector to normal plane

While in figure 5,

> | AE | is the distance from point P to the "normal plane";

> | BF | is the distance from point Q to "normal plane";

> In the same way, $fp\ (P) =$ | AE |, $fp\ (Q) =$ | BF |. And then,

∵ | AE | <| BF |,

∴ *fp (P) < fp (Q)*.

That is to say the abnormal probability of point P is less than that of point Q.

4.2 Synthetic Distance Model

The two above conclusions in figure (a) and figure (b) are inconsistent. In order to achieve coincidence of the result of the judgement, we make,

$$fp(P) = \frac{|AE|}{|AE| + |AC|}$$

$$fp(Q) = \frac{|BF|}{|BF| + |BD|}$$

Considering both the distance from "normal Plane" and the distance from "abnormal plane", we give a synthetic model shown in Figure 6.

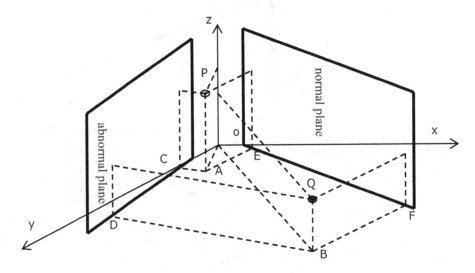

Fig. 6. The synthetic distance model of behavior vector

4.3 Synthetic Dimension Reduction Model

In accordance with the model proposed in above sections, we defined the normal degree, abnormal degree, the intrusion probability of a network behavior vector $\overrightarrow{X}$ as following.

Definition 3. Normal Degree

Let *AN* represents the normal behavior sample set in IDS database. Then we define the similarity between $\vec{X}$ and *AN* as S_{AN}^{X}, called the normal degree of the behavior vector $\vec{X}$. See formula (5).

$$S_{AN}^{X} = Sim(\vec{X}, AN) = \max\{Sim(\vec{X}, \overrightarrow{AN_j}), \overrightarrow{AN_j} \in AN, j = 0,1,\cdots,m\} \quad (5)$$

Definition 4: Abnormal Degree

Let *AI* presents the abnormal behavior sample set in IDS database. Then we define the similarity between $\vec{X}$ and *AI* as S_{AI}^{X}, called the abnormal degree of the behavior vector $\vec{X}$. See formula (6).

$$S_{AI}^{X} = Sim(\vec{X}, AI) = \max\{Sim(\vec{X}, \overrightarrow{AI_j}), \overrightarrow{AI_j} \in AI, j = 0,1,\cdots,m\} \quad (6)$$

Definition 5: Intrusion Probability

Taking both the normal degree and the abnormal degree of $\vec{X}$ into consider, we define the intrusion probability of $\vec{X}$ as $P(\vec{X}, AN, AI)$. It can be calculated by means of formula (7).

$$P(\vec{X}, AN, AI) = \frac{S_{AI}^{X}}{S_{AI}^{X} + \beta \bullet S_{AN}^{X}}, (\beta \geq 0) \quad (7)$$

In formula （7），

β is the contribution coefficient of the normal degree of $\vec{X}$ to the intrusion probability $P(\vec{X}, AN, AI)$. The value of β is not less than 0.

5 The Dimension Reduction Method of Behavior Samples

5.1 Mapping to 2-Dimension Points

According to formula (3) and formula (4), every network behavior $\vec{X}$ can be respectively result in a normal degree S_{AN}^{X} and anomaly degree S_{AI}^{X}. If let $u = S_{AN}^{X}$, $v = S_{AI}^{X}$, then, $\vec{X}$ can be mapped into a binary tuples <*u*, *v*>. Regard *u* as the longitudinal coordinates, and *v* as the abscissa, then (*u*, *v*) is a point in a 2-dimensional plane.

Taked notice of $u \in [0,1]$ and $v \in [0,1]$, a network behavior will be mapped into a point within the rectangle region from (0,0) to (1,1). This mapping is illustrated in Figure 4. Point (*u₁*, *v₁*) and (*u₂*, *v₂*) respectively represent the behavior vector $\vec{X_1}$ and $\vec{X_2}$.

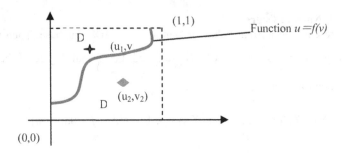

Fig.7. The mapping model from a behavior to a point

5.2 Behavior Detection Method

The final goal of a IDS is to alert a intrusion in time. In order to determine whether an behavior is abnormal or not, we need to define a threshold function $u = f(v)$. The curve of this function landed in the region of (0,0) to (1,1) while $v \in [0,1]$. This idea is illustrated in Figure 4 with a thick solid curve. Ideally, the curve of this function divide whole rectangle into two regions, D_1 and D_2. Observing from the figure, D_1 lies at the upper left side of the curve, and D_2 lies at the lower right corner. Point (u_1, v_1)fall in the region D_1, and (u_2, v_2) fell in the region D_2.

Let u_x denote u value of $\overrightarrow{X}$, and v_x denote v value of $\overrightarrow{X}$, then,

$$u_x < f(v_x) \tag{8}$$

It is to say that point (u_1, v_1) fall in D_1. (u_1, v_1) is not a intrusion. On the contrary, (u_2, v_2) fall in D_2. (u_2, v_2) is abnormal. Then, the IDS alarms in accordance with its security policy.

6 Conclusion

Aiming at the massive audit-data processing problem, which intrusion detection system facing at, we established a network behavior model in this paper. Then we calculated the normal probability and the abnormal probability to map a multi-dimension behavior vector to a 2-dimension point. According to the position of the 2-dimension point, IDS could recognize whether a behavior was a intrusion or not. This approach is easy to be parallel processed, and it will be very helpful to improve the intrusion detection efficiency in a high-speed distributed network.

Acknowledgements

This work is supported by Natural Science Foundation of Hebei under the Grant No.F2009000929 and National Natural Science Foundation of China under the granted No.60863003.

References

1. Jerne, N.K.: The immune system. J. Scientific American 229, 51–60 (1973)
2. Xiao, R.B., Wang, L.: Artificial Immune System: Principle, Models, Analysis and Perspectives. J. Chinese J. Computers 25, 1281–1293 (2002) (in Chinese)
3. Forres, S., Perelson, A.S., Allen, L., et al.: Self-Nonself discrimination in a computer. In: 1994 IEEE Symp. on Research in Security and Privacy, pp. 120–128. IEEE Press, New York (1994)
4. Kemmerer, R.A., Vigna, G.: Intrusion detection: a brief history and overview. J. IEEE Computer 35, 27–30 (2002)
5. Jiang, J.C., Ma, H.T., Ren, D.E.: Intrusion Detection Network Security: A Survey. J. Journal of Software 11, 1460–1407 (2000) (in Chinese)
6. Seung, H.S., Lee, D.D.: The Manifold Ways of Perception. J. Sience 290, 2268–2269 (2000)
7. Patcha, A., Park, J.M.: Network anomaly detection with incomplete audit data. J. Computer Networks 51, 3935–3955 (2007)
8. Fugate, M., Gattiker, J.R.: Anomaly detection enhanced classification in computer intrusion detection. In: Lee, S.-W., Verri, A. (eds.) SVM 2002. LNCS, vol. 2388, pp. 186–197. Springer, Heidelberg (2002)
9. Kim, D.S., Park, J.S.: Network-Based intrusion detection with support vector machines. In: Kahng, H.-K. (ed.) ICOIN 2003. LNCS, vol. 2662, pp. 747–756. Springer, Heidelberg (2003)
10. Park, J.S., Shazzad, K.M., Kim, D.S.: Toward modeling lightweight intrusion detection system through correlation-based hybrid feature selection. In: Feng, D., Lin, D., Yung, M. (eds.) CISC 2005. LNCS, vol. 3822, pp. 279–289. Springer, Heidelberg (2005)
11. Carol, T., Jim, A.F.: NATE: Network analysis of anomalous traffic events, a low-cost approach. In: 2001 Workshop on New Security Paradigms, pp. 89–96. ACM Press, New Mexico (2001)
12. Horng, S., Fan, P., Chou, Y., et al.: A feasible intrusion detector for recognizing IIS attacks based on neural networks. J. Computers & Security 27, 1–17 (2008)
13. Zhang, C.Y., Cao, Y.D., Yang, M.H., et al.: The Immune Recognition Method Based on Analogy Reasoning in IDS. J. Wuhai University Journal of Natural Sciences 11, 1839–1843 (2006)

Author Index